SCHOOL BOND SUCCESS

HOW TO ORDER THIS BOOK

BY PHONE: 800-233-9936 or 717-291-5609, 8AM–5PM Eastern Time

BY FAX: 717-295-4538

BY MAIL: Order Department
Technomic Publishing Company, Inc.
851 New Holland Avenue, Box 3535
Lancaster, PA 17604, U.S.A.

BY CREDIT CARD: American Express, VISA, MasterCard

BY WWW SITE: http://www.techpub.com

PERMISSION TO PHOTOCOPY–POLICY STATEMENT

Authorization to photocopy items for internal or personal use, or the internal or personal use of specific clients, is granted by Technomic Publishing Co., Inc. provided that the base fee of US $3.00 per copy, plus US $ .25 per page is paid directly to Copyright Clearance Center, 222 Rosewood Drive, Danvers, MA 01923, USA. For those organizations that have been granted a photocopy license by CCC, a separate system of payment has been arranged. The fee code for users of the Transactional Reporting Service is 1-56676/99 $5.00 + $ .25.

SCHOOL BOND SUCCESS

A Strategy for Building America's Schools

Floyd Boschee **Carleton R. Holt**

With Contributions from Patricia M. Peterson

Foreword by Paul D. Houston
Afterword by Senator Tom Daschle

LANCASTER · BASEL

LB
2825
.B63
1999

School Bond Success
a **TECHNOMIC** publication

Technomic Publishing Company, Inc.
851 New Holland Avenue, Box 3535
Lancaster, Pennsylvania 17604 U.S.A.

Copyright ©1999 by Technomic Publishing Company, Inc.
All rights reserved

No part of this publication may be reproduced, stored in a retrieval system, or transmitted, in any form or by any means, electronic, mechanical, photocopying, recording, or otherwise, without the prior written permission of the publisher.

Printed in the United States of America
10 9 8 7 6 5 4 3 2 1

Main entry under title:
School Bond Success: A Strategy for Building America's Schools

A Technomic Publishing Company book
Bibliography: p.
Includes index p. 165

Library of Congress Catalog Card No. 98-87282
ISBN No. 1-56676-705-9

To our wives and children,
Marlys Ann Boschee and Barbara, Brenda, Bonni, and Beth
Judy Holt and Austin and Cody

TABLE OF CONTENTS

FOREWORD

THERE are those who think that the best education is the simplest and that we should go back to the days when Socrates and Plato sat on a log and studied together. They say you do not need buildings in which to learn. You can learn with a teacher and a group of students sitting under a tree. This is true. However, it might get a bit uncomfortable in the rain or snow, and there is no place to plug in your computer. I often heard this expressed when my own school district needed to discuss building new schools or was looking to renovate the ones we had. Sadly, many adults in America seem perfectly content to let our children study in inadequate facilities that they, as adults, would not spend time in because of their condition.

Those folks need to remember that Socrates had to take hemlock because he was corrupting the youth of Athens. He is no longer around, and the style of education he offered, while good for Plato, is a bit dated and unrealistic in a country serving over 50 million school children in the Information Age.

We know, based upon studies done previously by my own organization, American Association of School Administrators, and later confirmed by the General Accounting Office, that we have over a 100 billion dollar problem in school facilities. This is because of a combination of deferred maintenance, student growth, and the changes needed because of new technology. And the problem continues to grow.

The reality is that school leaders have to constantly restate the case for better facilities for children. While it is self-evident to those working in schools that our buildings are inadequate and that more and better facilities are needed, it is not clear to the public, who must be convinced of the need and encouraged to pay for these costs.

Raising capital for new projects comes from the bonding process. It involves identifying the needs, creating a bond package, taking it to the voters, convincing them of the needs, and once approved, selling the bonds. Floyd Boschee and Carleton R. Holt have, in this publication, provided a comprehensive overview of the issue. In essence, they have given you all you need to know about creating school bond success. This book gives the school leader an excellent overview of the steps involved in creating a successful bond program, with tips, checklists, and insights on the details. If you are thinking about doing something about the facilities in your district, you should read this book. Socrates would have.

PAUL D. HOUSTON, ED.D.
Executive Director
American Association of School Administrators

PREFACE

NO one familiar with problems in education would fail to include the state of America's educational infrastructure as a primary concern. Many school buildings are falling apart due to age or lack of maintenance. Many school buildings do not meet the needs of modern educational programs and curricula. Unfortunately, school administrators and boards of education have found it increasingly difficult to obtain the funds necessary to correct facilities' problems in their districts. It is a problem that the educational community needs to address as a major obstacle to providing the high quality educational services that parents, students, and business people are demanding.

The higher education system must bear some of the responsibility for this problem. While many administration courses outline how to design school buildings, how to estimate school enrollments, and how to figure square footage requirements, there are few resources that inform top administrative officials about the nuts and bolts of getting the money. This book attempts to alleviate that gap.

School Bond Success: A Strategy for Building America's Schools includes the theoretical basis for developing specific strategies for resolving school bond issues and practical information on specific activities for bond campaigns. Chapter 1 provides the case for the need and information from the research as to essentials in successful bond campaign strategies. Chapters 2 through 7 provide step-by-step instructions for the four stages of planning building projects: the pre-planning phase, the project development phase, the bond campaign phase, and the follow-up phase. Chapter 8 addresses particular issues that may cause concern to volunteers and others who are involved in conducting the bond campaign, including conducting surveys, dealing with conflict, and writing proposals and campaign materials. The appendices provide the reader with a checklist for effective school bond campaigns, planning in debt issuance, and a glossary of terms as they relate to municipal bonds.

Each chapter also includes specific review activities. These exercises, when followed in order, allow the person using this book to develop specific materials for their school districts as they proceed through the bond process. The ideas in this book can be used in all types of school districts: from large, urban districts to small, rural districts.

Some may say that the ideas presented are just "common sense"; however, the concepts presented have been tried and tested and have been shown to be effective. By using the techniques offered in this volume, administrators do not have to "reinvent the wheel" or experiment (sometimes with disastrous results).

Providing appropriate and safe environments in which students may take advantage of educational opportunities should be one of the top priorities of the educational community. Developing effective strategies for accomplishing that task should be one of any school administrator's major goals.

FLOYD BOSCHEE
CARLETON R. HOLT

ACKNOWLEDGMENTS

NO single writer or pair of writers could prepare a comprehensive work on school bond success without relying heavily upon ideas, prior publications, and constructive criticisms of others. We appreciate the cooperation of those whose published writings, research, and private documents are quoted in this volume. We are equally grateful to all those whose writings are cited as references.

We would like to express our sincere appreciation to Dr. Marlene J. Lang, Instructional Design Specialist, University of South Dakota, Vermillion, for her guidance and editorial advice. Gratitude is extended to Patricia M. Peterson for the many contributions. Her translation of the rough draft to the finished product helped make this book possible. A special thank you is extended to Darwin Reider from Kirkpatrick Pettis, Omaha, NE; Todd Meierhenry from Danforth, Meierhenry, & Meierhenry, L.L.P., Sioux Falls, SD; and the Board of Education from the Brandon Valley School District, Brandon, SD, for providing information and granting permission to use their material for this book.

During the preparation of Appendix A, State School Bond Referendum Percentages, we relied upon library searches conducted in the McKusick Law Library by graduate students in the School Law class at the University of South Dakota. Matt Beckendorf, Jane Bradfield, Kevin Brick, Tina Cameron, Katherine Campbell, Jerry Carda, Lance Hankins, Darrell Langely, Trevor Long, Joan Mahoney, Christina Nelsen, Corey Peters, Tim Pflanz, Rustin Roland, Lisa Stoebner, and Philip Willenbrock applied superb library skills as information was gathered for the fifty states cited in Appendix A.

We dedicate this book to school administrators, school board members, and community members who understand how valuable the school learning environment is and how it can affect the lives of young people. The efforts put forth to have adequate school buildings by the educational leaders make it possible for children to be properly educated and prepared for the future.

CHAPTER 1

The Problem with America's Aging Schools

PUBLIC school facilities for elementary and secondary school students in the United States, almost exclusively a state and local responsibility, should consist of infrastructures that create an environment in which all children can receive a proper education and preparation for the future. There is, however, a growing concern about the safety and adequacy of many of this country's school buildings. "Recently, for example, a judge would not allow the schools in the nation's capital to open on time until thousands of life-threatening fire code violations were corrected."[1] Similarly, over 1,000 New York City schools had to close for eleven days because of noncompliance with asbestos requirements.[2] These are but two examples of a problem that is facing countless school districts across the country: how to provide school facilities that meet the educational and health needs of America's millions of students.

THE PRESENT STATE OF AMERICA'S SCHOOLS

Elementary and secondary education, the largest public enterprise in the United States, is conducted in over 80,000 school buildings in about 15,000 school districts serving over 41 million students. Roughly 70 percent of the school buildings serve about 27 million elementary students, 24 percent serve about 13.8 million secondary students, and six percent serve approximately 1.2 million students in combined elementary and secondary and other schools.[3] Nationwide, about two-thirds of the school facilities are in adequate (or better) condition, needing at most some preventive maintenance or corrective repair. The

Table 1.1. Schools and students with less-than-adequate physical conditions.

Building Feature	Number of Schools	Students Affected
Heating, ventilation, air conditioning	28,100	15,456,000
Plumbing	23,100	12,254,000
Roofs	21,000	11,916,000
Exterior wall, finishes, windows, doors	20,500	11,524,000
Electrical power	20,500	11,034,000
Electrical lighting	19,500	10,837,000
Interior finishes, trims	18,600	10,408,000
Life safety codes	14,500	7,630,000
Framing, floors, foundations	13,900	7,247,000

Source: GAO/HEHS-95-61. February 1995. *School Facilities,* p. 10.

remaining one-third of both elementary and secondary schools (25,000 schools and 14 million students), however, needs extensive repair or replacement of one or more buildings.[4]

More than seven million students must attend school in buildings that fail to comply with life safety codes (e.g., stairwells, adequate exits, panic hardware, fire extinguishers, rated corridor doors, fire walls, sprinkler systems, and the like); more than fifteen million students attend schools that lack adequate heating, ventilation and air conditioning; and over eleven million students frequent schools that need extensive roof repair[5] (see Tables 1.1 and 1.2).

The following examples provide sufficient evidence of this nationwide problem:

- In New Orleans, the damage from Formosan termites has deteriorated the structure of many schools. In one elementary

Table 1.2. Schools and students with unsatisfactory environmental conditions.

Environmental Condition	Number of Schools	Number of Students Affected
Ventilation	21,100	11,559,000
Acoustics for noise	21,900	11,044,000
Physical security	18,900	10,638,000
Indoor air quality	15,000	8,353,000
Heating	15,000	7,888,000
Lighting	12,200	6,682,000

Source: GAO/HEHS-95-61. February 1995. *School Facilities*, p. 10.

school, the insects ate both the books on the library shelves and the shelves themselves. This, in combination with a leaking roof and rusted window wells, caused so much damage that officials condemned a portion of the 30-year-old school.[6]

- At a Montgomery County, Alabama, elementary school, a ceiling weakened by leaking water collapsed 40 minutes after the children left for the day.[7]
- Water damage from an old (original) boiler steam heating system at a 60-year-old junior high school in Washington, D.C., has caused such wall deterioration that an entire wing has been condemned and locked. Steam damage is also causing lead-based paint to peel.[8]
- Raw sewage back[ed] up on the front lawn of a Montgomery County, Alabama, junior high school because of a defective plumbing system.[9]
- A New York City high school, built around the turn of the century, has served as a stable, fire house, factory, and office building. The school is overcrowded with 580 students, far exceeding the building's 400 student capacity. The building has little ventilation (no vents or blowers), despite many inside classrooms, and the windows cannot be opened, which makes the school unbearably hot in the summer. In the winter, heating depends on a fireman's stoking the coal furnace by hand.[10]
- In Ramona, California, where overcrowding is considered a problem, one elementary school is composed entirely of portable buildings. It has neither a cafeteria nor an auditorium and uses a single, relocatable room as a library, computer lab, music room, and art room.[11]
- During a windstorm in Raymond, Washington, the original windows of an elementary school built in 1925 were blown out, leaving shards of glass stuck in the floor. The children happened to be at the other end of the room. This wooden school is considered a fire hazard, and although hallways and staircases can act as chimneys for smoke and fire, the second floor has only one external exit.[12]
- In rural Grandview, Washington, overcrowded facilities are a problem. At one middle school, the original building was meant to house 450 students. Two additions and three portables have been added to accommodate 700 students. The school has seven

staggered lunch periods. The portables have no lockers or bathrooms and are cold in the winter and hot in the spring/summer.[13]

- In a high school in Chicago, the classroom floors are in terrible condition. Not only are floors buckling, so much tile is loose that students cannot walk in all parts of the school. The stairs are in poor condition and been have cited for safety violations. An outside door has been chained for 3 years to prevent students from falling on broken outside steps. Peeling paint has been cited as a fire hazard. Heating problems result in some rooms having no heat while other rooms are too warm. Leaks in the science lab caused by plumbing problems prevent the classes from doing experiments. Guards patrol the outside doors, and all students and visitors must walk through metal detectors before entering the school.[14]

One could dismiss the above examples as "extreme"; unfortunately, they are not. The conditions in facilities for one-third of the students attending public schools in the United States range from uncomfortable to downright dangerous. As illustrated in Table 1.3, unsafe and unhealthy conditions exist in all types of school buildings in every state in the country.

The consequences of failing to improve the infrastructures in America's school districts are too great to be ignored. Some of the effects have a direct impact on the educational mission of the schools. A study comparing student performance in a modern school building with student performance in an older school building revealed that "students in the modern building scored significantly higher in reading, listening, language, and arithmetic than did students in the older structure."[15] Further, "students in the modern school facility received significantly less discipline, had a significantly higher attendance record, and were in better health than students in the older buildings."[16] At the same time, these aging structures are not properly equipped to allow for many of the new technologies that should be utilized.

Attending school in older buildings also poses the possibility of long-term health problems for students. Children in those schools are exposed to hazardous substances such as asbestos, lead in water or paint, materials contained in underground storage tanks (UST), and radon—all substances directly linked to increasing incidences of cancer

Table 1.3. Estimated percent of schools with at least one building in inadequate condition by state.

State	Percent of Schools Reporting at Least One Inadequate Original Building	Percent of Schools Reporting at Least One Inadequate Attached and/or Detached Permanent Addition	Percent of Schools Reporting at Least One Inadequate Temporary Building	Percent of Schools Reporting at Least One Inadequate On-Site Building
Alabama	32.5	19.1	31.5	39.1
Alaska	36.7	21.7	22.8	44.6
Arizona	27.1	14.2	28.8	40.8
Arkansas	16.8	11.8	14.5	24.9
California	31.8	14.3	24.3	42.9
Colorado	21.3	12.3	16.5	32.2
Connecticut	27.1	13.7	8.0	30.0
Delaware	30.0	7.7	35.5	40.5
District of Columbia	49.3	20.7	0.0	49.3
Florida	18.3	10.7	20.9	31.2
Georgia	18.5	9.0	15.1	26.2
Hawaii	16.3	5.5	11.2	21.4
Idaho	27.4	14.9	13.3	31.9
Illinois	29.2	8.8	4.4	31.0
Indiana	28.1	11.5	2.6	29.2
Iowa	14.9	7.6	8.5	18.8
Kansas	33.7	14.5	18.8	38.3
Kentucky	24.0	12.9	17.7	30.9
Louisiana	28.0	8.7	24.8	38.6
Maine	34.5	14.5	13.0	37.5
Maryland	27.3	9.3	6.1	30.7
Massachusetts	37.8	11.8	4.9	40.8

continued

Table 1.3. (continued).

State	Percent of Schools Reporting at Least One Inadequate Original Building	Percent of Schools Reporting at Least One Inadequate Attached and/or Detached Permanent Addition	Percent of Schools Reporting at Least One Inadequate Temporary Building	Percent of Schools Reporting at Least One Inadequate On-Site Building
Michigan	19.4	9.9	4.9	21.6
Minnesota	32.8	16.9	16.4	38.5
Mississippi	14.5	9.6	19.1	28.5
Missouri	24.0	3.8	11.7	27.3
Montana	16.5	7.9	7.9	20.4
Nebraska	29.5	9.7	6.4	35.2
Nevada	20.9	4.6	10.1	23.2
New Hampshire	33.4	4.6	16.0	38.4
New Jersey	17.3	12.8	1.1	19.1
New Mexico	25.6	13.7	13.6	29.9
New York	28.6	8.5	5.7	32.8
North Carolina	25.0	9.6	24.5	36.1
North Dakota	20.5	10.0	6.7	23.0
Ohio	33.0	20.2	8.2	38.0
Oklahoma	27.1	11.3	16.0	30.5
Oregon	31.4	19.8	11.1	38.9
Pennsylvania	18.9	9.6	4.9	21.0
Rhode Island	29.3	13.8	0.0	29.3
South Carolina	21.2	13.6	29.4	36.9
South Dakota	20.1	12.0	8.4	21.3
Tennessee	18.6	10.6	14.0	27.2
Texas	22.6	13.2	13.2	27.1

Table 1.3. (continued).

State	Percent of Schools Reporting at Least One Inadequate Original Building	Percent of Schools Reporting at Least One Inadequate Attached and/or Detached Permanent Addition	Percent of Schools Reporting at Least One Inadequate Temporary Building	Percent of Schools Reporting at Least One Inadequate On-Site Building
Utah	34.4	22.0	3.4	34.1
Vermont	18.6	13.9	18.0	21.4
Virginia	20.8	16.1	10.8	27.4
Washington	37.6	16.9	25.2	44.2
West Virginia	39.5	25.3	15.8	41.9
Wisconsin	31.8	16.1	4.9	32.8
Wyoming	18.3	6.3	10.5	24.4

Source: GAO/HEHS-96-103, *School Conditions Vary*, pp. 33–34.

and other illness. Medical researchers have also linked poor ventilation in buildings to an alarming increase in incidences of respiratory problems and asthma over the past decade. A recent update in *American School & University* on legislation affecting education facilities revealed that a fifth-grade teacher suffering from severe allergies to mold filed a federal discrimination lawsuit under the Americans with Disabilities Act (ADA) against the Mineral County School District, W.V., claiming it failed to transfer her out of a classroom that made her gravely ill. Poor ventilation, inefficient heating and cooling systems, and a leaky roof were cited as causing the classroom's poor indoor air quality (IAQ).[17] While some older schools are structurally sound, others are accidents waiting to happen. At Raymond, Washington, all three schools are old and two may be unsafe. The high school, built in 1925, is a three-story structure of unreinforced concrete that may not safely withstand the possible earthquakes in the area. The elementary school, built of wood in 1924, is a potential fire hazard and will be closed in two years.[18]

In Hartford, South Dakota, "a 10-foot-tall chunk of chimney toppled from the building, crashed through the roof and a classroom, and landed in the first-floor library"[19] (see Photo 1.1). Fortunately, the school was closed for a holiday. The superintendent indicated that the problem was with aging bricks and mortar, a problem that probably could not have been detected by an inspection.[20]

These "near misses" indicate that school districts that continue to hold classes in facilities that have been in use beyond their "life expectancy" are running a tremendous risk. Not only would these two incidents have been tragedies if students and teachers had been in the affected areas, the liability of the school district probably would have been greater than the cost of building a new school.

School budgets also suffer when buildings are in a state of disrepair. Most unrepaired buildings are not energy efficient, and school boards unnecessarily spend money on wasted fuel and electricity that they could use for other programs. One elementary principal summed up the problem well: "Heat escapes through holes in the roof; the windows leak (ones that are not boarded up) and let in cold air in the winter so that children must wear coats to class."[21]

Still other schools have delayed maintenance because of the lack of funds. Deferred maintenance, however, speeds up the deterioration of buildings. In an elementary school in New York City, repair problems

***Photo 1.1**. School averts tragedy when chimney tumbles, West Central Junior High School, Hartford, SD. (Photo courtesy of the* Argus Leader, *Sioux Falls, SD. Used by permission.)*

have not been addressed since the school was built twenty years ago. Years ago, the problems could have been addressed relatively inexpensively; now the ongoing lack of attention has caused sewage to leak into the first-grade classrooms and a leaking roof has caused the building to become structurally unsound.[22]

In Chicago, an elementary school roof had needed replacement for twenty years. The roof had been superficially patched rather than replaced, and the persistent water damage caused floors to buckle and plaster on the walls and ceilings to crumble. Parts of the electric wiring system also had been continuously flooded. One teacher in the school would not turn on the classroom lights during rainstorms for fear of electrical shock. In another classroom the public address system was rendered unusable. Janitors had to place buckets on the top floor of the school to catch the rain.[23]

That a problem exists is clear. That students and educators are being harmed by the failure of school districts to repair or replace inadequate structures also is clear. What is not so clear is how to solve the problem.

OBSTACLES TO FACILITIES DEVELOPMENT

Of course, the largest obstacle for school boards in their attempts to provide adequate school facilities is money—or more correctly, the lack thereof. Researchers estimate "that the nation's schools need 112 billion dollars to complete all repairs, renovations, and modernizations required to restore facilities to good overall condition and to comply with federal mandates"[24] (see Figure 1.1). One-third of the schools that remain unsafe need 65 billion dollars ($2.8 million per school) for major repairs or replacement. Forty percent of the schools that are now "adequate" need 36 billion dollars ($1.2 million per school) to repair or replace one or more building features (e.g., roofs; framing, floors, and foundations; exterior walls, finishes, windows, and doors; interior finishes and trims; plumbing, heating, ventilation, and air conditioning; electrical power; electrical lighting; and life safety codes).[25]

Nearly two-thirds of the schools need roughly 11 billion dollars (an average of $.2 million per school) to comply with federal mandates over the next three years. From those 11 billion dollars, schools need approximately five billion dollars (54%) to correct or remove hazard-

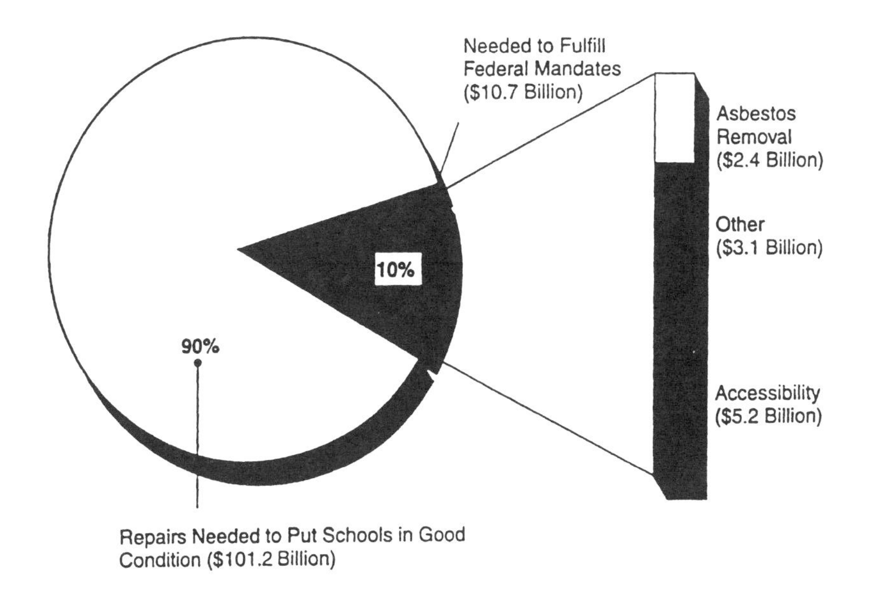

Note: "Other" includes lead in water/paint, underground storage tanks, radon, and other mandated requirements.

Figure 1.1. *Dollars needed for repairs to comply with federal mandates (*Source: *GAO/HEHS-95-61, p.6).*

ous substances, such as asbestos, lead in water or paint, materials contained in UST, and radon or meet other requirements while over 6 billion dollars (55%) is needed to make programs physically accessible to all students.[26]

Some people may view the 112 billion dollar figure as outrageously high. An examination of just one state, however, illustrates the magnitude of the problem across the country. South Dakota is a state with one of the lowest populations in the nation. A preliminary study[27] of the square footage for school use in the state's 172 school districts (two districts do not have buildings) revealed that 13 percent (2,700,000 square feet) of the square footage in use is over 70 years old, 9.8 percent (2,035,000 square feet) is between 51 and 70 years old, 29.9 percent is between 31 to 50 years old, 29.1 percent (6,230,000 square feet) is between 11 and 30 years old, and 17.4 percent (3,615,000 square feet) of the buildings are from 0 to 10 years old. The portable units comprise 0.8 percent (165,000 square feet) of the square footage in use. The price tag for making all public school facilities in South Da-

Table 1.4. What a new school costs.

	Elementary School	Middle School	High School
Cost/Square Foot	$110.57	$108.59	$109.09
Cost/Student	$12,667	$15,143	$18,885
Square Feet/Student	114	150	160
Average No. Pupils	600	800	900
Average Size (sq. ft.)	60,000	114,000	133,500
No. Classrooms	26	38	35
Total Cost	$7,112,917	$12,228,198	$16,031,336

Source: American School & University's 23rd Annual Official Education Construction Report, May 1997. Used by permission.

kota adequate, accessible, and environmentally acceptable is 434 million dollars (nearly one million dollars per building).[28]

If the needed expenditures in a low-population state such as South Dakota exceed 400 million dollars, logic would indicate that the suggested 112 billion dollar figure may not be sufficient to repair or replace all of the inadequate school structures in the nation. *The American School & University's 23rd Annual Education Construction Report,* May 1997, disclosed that the average elementary school for 600 students costs about $7.1 million; a middle school for 800 students, $12.2 million; and a high school for 900 students, $16 million (see Table 1.4).[29] And these figures do not take into account the need to update existing facilities to accommodate new technologies, enlarge spaces for special services, and provide flexible learning environments for new instructional programs.

For many school districts already struggling to pay for "essential" budget items, such repair and building costs cannot come out of the usual tax revenues, and in many cases opposition to such projects is too great to overcome.

CHANGING ENROLLMENTS

For the past decade, enrollments in many elementary and secondary schools have declined. Most school officials did not have to concern themselves with the possibility of having to build new schools to handle the students; indeed, in many instances the problem was determin-

Table 1.5. Percent change in grades K–1 enrollment in public schools by region and state with projections: Fall 1987 to fall 2005.

Region and State	Actual	Projected		
	1987 to 1993	1993 to 1999	1999 to 2005	1993 to 2005
United States	8.4	10.2	3.6	14.5
Northeast	5.5	8.8	0.5	9.4
Connecticut	7.5	8.9	0.6	9.6
Maine	3.0	0.7	−3.2	−2.5
Massachusetts	5.9	9.5	−1.4	8.0
New Hampshire	12.0	7.9	−1.2	6.6
New Jersey	5.0	13.6	5.2	19.5
New York	5.0	8.6	0.9	9.6
Pennsylvania	5.3	7.0	−1.5	5.5
Rhode Island	8.7	7.4	−1.4	5.9
Vermont	7.4	5.4	−0.5	4.9
Midwest	4.2	6.1	0.7	6.9
Illinois	4.2	7.8	2.6	10.6
Indiana	0.1	6.3	3.4	9.9
Iowa	3.7	2.1	−2.4	−0.3
Kansas	8.6	6.5	2.0	8.6
Michigan	1.9	7.3	2.2	9.7
Minnesota	12.3	6.6	−2.4	4.1
Missouri	7.3	4.8	−0.3	4.5
Nebraska	5.9	4.5	−0.2	4.3
North Dakota	−0.9	−2.7	−2.4	−5.0
Ohio	0.2	5.0	0.3	5.3
South Dakota	7.4	5.7	0.9	6.7
Wisconsin	10.0	6.4	−2.5	3.7
South	7.5	10.2	4.1	14.8
Alabama	−0.9	8.8	8.3	17.8
Arkansas	1.4	5.3	2.7	8.2
Delaware	11.5	13.5	4.2	18.2
District of Columbia	−8.6	−7.7	−1.0	−8.6
Florida	22.7	15.6	1.7	17.5
Georgia	10.9	13.2	5.5	19.4
Kentucky	1.5	3.0	1.7	4.7
Louisiana	−1.0	1.4	2.4	3.8
Maryland	12.5	16.8	6.1	23.9
Mississippi	−1.0	1.5	1.9	3.3
North Carolina	3.7	14.2	5.6	20.7
Oklahoma	3.1	4.0	0.6	4.6
South Carolina	3.8	8.9	4.5	13.7
Tennessee	3.6	9.3	3.3	12.9
Texas	10.7	10.5	5.5	16.6
Virginia	7.5	14.1	4.9	19.8
West Virginia	−8.8	−2.3	0.8	−1.5

(continued)

Table 1.5. (continued).

Region and State	Actual 1987 to 1993	Projected 1993 to 1999	Projected 1999 to 2005	Projected 1993 to 2005
United States	8.4	10.2	3.6	14.5
West	17.1	16.9	7.8	26.1
Alaska	18.5	18.0	5.9	25.0
Arizona	21.3	17.5	5.2	23.6
California	18.7	19.1	9.2	30.1
Colorado	11.6	14.6	3.2	18.3
Hawaii	7.9	15.8	11.9	29.6
Idaho	10.2	10.3	7.6	18.6
Montana	6.4	5.4	1.8	7.3
Nevada	39.2	24.0	3.9	28.9
New Mexico	10.9	15.0	8.3	24.6
Oregon	14.2	12.6	6.3	19.6
Utah	10.4	6.4	8.2	15.2
Washington	18.8	17.6	6.7	25.5
Wyoming	1.7	2.0	6.8	9.0

Source: U.S. Department of Education, National Center for Education Statistics. Common Core of Data Surveys and "Public Elementary and Secondary Education Statistics Year 1993–1994." *Early Estimates.*

ing which school to close. The National Center for Education Statistics (NCES) predicts, however, that public school enrollments in the United States will increase over the next decade, albeit such growth will vary across the country (see Table 1.5).[30] The Northeast will have enrollment increases in most states between 1993 and 1999, and between 1999 and 2005 most states will see slight declines. Increases for this region range from a 20 percent increase in New Jersey to a five percent increase in Vermont. Only Maine will show a decrease.

Increases in enrollments will vary more across the Midwest, but increases should continue in most states until 2005. The NCES predicts the highest increase in Illinois (11%); they predict the lowest increase in Wisconsin (4%). The NCES predicts that North Dakota and Iowa will experience small declines.

In the South, researchers predict that increasing enrollments should continue in most states until 2005. The greatest increase should occur in Alabama (18%) and the smallest increase should occur in Mississippi (3%). NCES estimates that the District of Columbia will have a sizable decrease in enrollments (9%) and that West Virginia will have a slight decrease (1%).

The NCES expects all states in the West to show enrollment increases from a startling 30 percent increase in California to smaller increases in Montana (7%) and Wyoming (9%). The number of school children in these states is expected to increase through 2005.

The above changes in enrollment set up unusual challenges for school administrators across the country in terms of making decisions about existing and needed buildings within their school districts. If enrollments are due to increase over the next few years and then decline, administrators must determine whether it is most advantageous to build or to wait it out. The situation also creates a dilemma for administrators who must determine whether to divert monies needed for repair or restoration of existing structures to pay for new buildings that would alleviate overcrowding.

FACTORS CONTRIBUTING TO THE DECLINE IN THE CONDITIONS OF SCHOOL FACILITIES

Clearly, administrators and boards of education face many dilemmas in terms of their school districts' physical plants. Many simply do not have the necessary money available from their general and/or capital outlay funds to solve their building problems. Additionally, they often face attitudes within their communities that become barriers to taking action.

A lack of adequate funding of education from property taxes has resulted in the delayed maintenance and repair of some buildings and the delayed replacement of aging and outdated facilities in many school districts. Over the past several years, a formidable anti-tax movement has developed across the country. In many states, such movements have brought about legislated tax limitations and restrictions on capital expenditures. Such measures often place severe pressure on already strained school budgets. No one likes to pay more taxes than necessary, but these movements may indicate underlying negative views of education that can spell disaster for bond referendum attempts. (About one in three school districts have had an average of two bond referendums fail in the past two years.)

Changing societal demographics also may affect the outcomes of elections. As the population ages, fewer citizens in any given community have school-age children. Older residents sometimes believe that existing structures "were good enough for me," and therefore reason that such structures are good enough for today's students. Many citizens also believe that schools spend too much money on such programs as special education that are federally and state-mandated and require considerable funding. District patrons do not like having such programs forced upon them and sometimes resent having to pay for them. When the schools need money for equipment or buildings, these individuals may form strong opposition groups.

Often school budgets are stretched to their limits by the imposition of both unfunded federal- and state-mandated programs. Such expenditures cause the diversion of discretionary monies that school administrators could use for repair and replacement of outdated structures.

To deal effectively with this problem, school administrators and boards of education must begin by reviewing their alternatives in terms of financing the needed repairs and replacements. Then, they must realistically assess their building needs and begin developing long-range plans for repairing and replacing inadequate facilities.

FINANCING SCHOOL BUILDINGS

With the advent of various state grants to local school districts, complete responsibility for financing public schools was eliminated in the early 1900s. However, the tradition of local responsibility for capital-outlay costs has not changed. "During the many years of almost exclusive local support for financing capital outlays, several different plans and procedures have evolved in various states. Chief among these have been pay-as-you-go plans, use of reserve funds, and bonding."[31]

Pay-as-You-Go Financing

School districts that utilize "pay-as-you-go financing" pay for construction of school facilities from current revenues. The school district pays the entire cost of a building project from the receipts of one fiscal year's local tax levy. This method proves ideal because it is the quickest and easiest way to finance capital outlay projects. The system also eliminates payment "of large sums of money for interest, which can represent from 20 to 40 percent of the cost of a building, the costs of bond attorney fees, and election costs."[32] Unfortunately, this alternative is available only to large and/or very affluent districts.

Building Reserve Funds

Where such action is legal, school districts often accumulate taxes from one year to the next for the purpose of constructing future school facilities. "This plan provides for spreading construction costs over a period of time before the buildings are erected—as contrasted with bonding, which spreads the cost over time after the schools are constructed."[33] The benefits of this system to a local community are great. For example, buildings can be constructed without delay, debt service charges and expenses associated with attaining voter approval for issuance of bonds are eliminated and local restrictions on taxing or debt limitation will not interfere with the building project.[34] Here again, state laws and regulations may limit the number of school districts that can utilize this plan.

Bonding

Bonding is the most common local program for financing public school facilities. "The process involves obtaining taxpayer favor for the [school] district to issue long-term bonds to obtain funds to construct buildings and provide other facilities."[35] School districts with low assessed valuations of property, insufficient tax revenues to finance building costs on a current basis, and an inability to accumulate reserve funds use bonding practices.

State governments often restrict the ability of school districts to assess taxes for use in capital outlay projects, and many states determine the types of projects for which such monies can be used. In most states, schools maintain a capital outlay fund to meet expenditures exceeding three hundred dollars that result in the acquisition of or additions to real property, plant, or equipment. In nearly all states such expenditures are limited to the purchase of land, renovation of existing facilities, improvement of grounds, construction of new facilities, additions to facilities, remodeling of facilities, or the purchase of equipment. It may also be used for installment or lease-purchase payments for the purchase of real property, plant, or equipment.[36]

Additionally, most states limit the amount school districts may go into debt for such expenditures. For example, in South Dakota the amount cannot exceed three percent of the taxable valuation. In Colorado, non-unified school districts cannot exceed 1.25 percent of taxable property and unified school districts cannot exceed 2.5 percent.[37]

Most states also control the dates of elections, who conducts the elections, repayment requirements, and the percent by which such elections must prevail in order to be passed. The superintendent and school attorney should check their own state laws for this information. (For more information on the bonding process, see Chapter 3.)

THE PROBLEM WITH BOND REFERENDUMS

While school administrators may use any of the three alternatives listed above to fund capital outlay projects, most school districts utilize the bonding process for major building or repair projects simply because it is the only way they can obtain the necessary money. But securing funds through this system is not an easy task. The school districts in most states must hold special elections, and states often require that the issue pass by more than a simple majority vote (see Appendix A for percentage needed in each state to pass a school bond referendum). For example, in Mississippi, three-fifths of the votes cast must be favorable for the issue to pass and in California, two-thirds of the votes must be favorable. Moreover, if boards of education place their school district residents in debt, they must be able to validate reliability. States that require a 60 percent or more voter approval trumpet Condorcet's time-tested theory of group reliability (see Exhibit 1.1). For school districts that have fewer than ten thousand eligible voters, the 60 percent vote is required to obtain a 99.97 percent of reliability.[38] An analysis of school bond elections in California, which requires a two-thirds majority for passage, since 1982 reveals that 50.47 percent of those elections failed, but 71.27 percent would have passed with a 60 percent, 75.99 percent would have passed with a 58 percent, and 93.01 percent would have passed with a 50 percent + 1 threshold. Iowa requires a 60 percent majority for passage. From 1990 to 1995, the Iowa State Department reported that only 38 percent of the Iowa school districts holding bond referendums passed. The super majority concept clearly represents a factor in why school bond issues are so difficult to pass.[39]

The problem of passing school bond referendums is exacerbated

$$\frac{V^{h-k}}{V^{h-k} + E^{h-k}}$$

Assumed that each voter is right in **V** of the cases, and wrong in **E** of the cases (V + E = 1), and **h** voters vote yes, while **k** voters vote no, the probability that the **h** members are right is given by the above formula. The average voter is more often right than wrong.

Voter Reliability

# of Voters	% Required to Carry	% Individual Reliability	% Group Reliability
1,000	50 + 1	50	69.00
10,000	50 + 1	50	99.97
1,000	**60**	50	99.97

Source: Adapted from Black, D. 1987. The Theory of Committees and Elections. Boston: Kluwer Academic Publishers, pp. 159–185. Used by permission.

Exhibit 1.1. *Condorcet's reliability formula.*

by the fact that few institutions of higher education prepare school administrators for the task. Most of the textbooks on educational facilities used to prepare school administrators are directed toward the specifics of planning educational facilities for the future. They adequately address the historic development of educational facilities, how to determine school building needs, how to plan a building, how to modernize a building, and how to finance the capital outlay. But few, if any, provide specific strategies on how to win bond referendums. In many instances, school bond issues have failed because administrators were not prepared to plan an effective strategy for passing them.

STRATEGIES FOR SUCCESS

To plan a successful school bond referendum, school officials should be aware of those factors that influence election outcomes. Researchers and school officials from all parts of the country have identified several factors that contributed to either the success or failure of school bond elections. The following is a compilation of information obtained from more than fifty references.[40]

Factors That Influence Campaigns

Citizen Participation

Several reports indicate that the most significant factor in passing a school bond referendum is the development of a broad-based citizens' volunteer committee. This group appears to be effective when its members take the leadership roles in the campaigns. The responsibilities of these committees include gathering preliminary information about community attitudes through telephone surveying, conducting "get-out-the-vote" drives, and making presentations to community groups.

Community Relations Programs

The researchers also found that an ongoing community relations program that functions as a regular part of school district services is essential. The most important aspects of this service include providing information about the needs of the school and providing special services to senior citizens.

Consultants

Most school districts that were successful in bond elections found that utilizing consultants in the areas of school planning and finance was important. Most lay persons, including school administrators, do not have the knowledge, nor do they have the time to obtain the knowledge necessary to answer many of the questions that might arise in the course of discussions of the referendum. Consultants can provide appropriate and accurate information to the voters—an important component to developing needed credibility with the public.

Unity of Purpose

Administrators who have passed bond issues successfully also point to the importance of unity between school staff and the board of education. If these groups send mixed messages to the public about the architectural design or the need for the bond issue, most voters will develop a concern about the advisability of the project. Voters apply the adage: "When in doubt, vote no."

Voter Turnout

A high voter turnout is another critical factor in achieving victory at the polls. This points to the need for a "get-out-the-vote" effort. Most researchers, however, warn against relying on absentee voters to increase voter participation. As well, many campaign coordinators point to the special attention their committees gave to encouraging "yes" voters to vote as a positive factor in their victories.

Endorsements

Endorsements from influential community figures appear to be helpful in passing school bond referendums; however, the involvement of political figures, excluding city council members, does not increase the potential for a school bond victory.

Opposition Groups

While opposition groups may be difficult to identify, a successful campaign often hinges on whether or not those working for the bond issue clearly understand the problems those in opposition to the bond referendum are expressing to other voters. Planners should make sure that presentations and media messages address these issues as being important and that the proponent representatives are prepared to deal with the issues.

Tax Increase Limitations

The amount of the increase in tax levies has a great influence on the outcome of bond elections. Researchers report that community support appears to drop drastically when the levy exceeds $2.00 per $1,000 valuation.

Timing and Length of Campaigns

Some literature indicates that the timing and length of campaigns can be a factor, but most researchers find that neither the time of year nor the length of the campaign were significant in the outcomes of the elections.

While the aforementioned factors are general in scope, they do provide insights into how to plan, implement, and conclude a successful school bond issue campaign. A comprehensive study[41] of school districts that had both won and lost school bond elections provides an even greater understanding of those variables (specific activities, etc.) that influence election outcomes. The study included an examination of responses from members of citizen support groups, newspaper editors, school administrators, and bankers in communities that both approved and defeated bond referendums. The individuals were asked to identify specifics about the proposed building project, activities that were beneficial, activities that were not beneficial, the nature of the oppositions' complaints, the type of media support the issue was given, and the amount of citizen participation in the campaign.

Variables That Contribute to the Success of School Bond Referendums

Voter Turnout

The respondents indicated that from 35.5 percent to 40 percent of the registered voters within a school district participated in the school bond referendums. Many of the respondents implied that making a special effort to get "yes" voters to the polls was critical to their success.

Citizens Support Group

Consistent with national data, the respondents indicated that an active citizens support group was critical to the success of their bond referendum campaigns. Those school districts that experienced the greatest success were those in which administrators played a "low-key" role, while members of the support group assumed the primary responsibility for educating the public.

Support group members also indicated which activities they considered the most important. They included: identification of the needs of the school district, leadership in the promotion of the bond issue, fund-raising, publicity, door-to-door and telephone canvassing of the community, developing a campaign theme, designing brochures, making presentations to community groups, conducting building tours, and participating in media events.

The study's respondents indicated that, to be successful, the support group must be made up of individuals from all segments of the community. Such inclusionary practices give more people a personal stake in the results of the election. Those from school districts with the greatest margins of victory suggest that fund-raising activities provide a good way of identifying persons who support the bond issue.

Media

Almost all participants in the study indicated that media support was important to the success of their election. Newspapers in communities where the bond issue passed carried editorials supporting the bond issue proposals and published letters to the editor supporting the bond issue proposals. Letters to the editor appear to have been an important component of informing the public as to the need for the raise in taxes. Letters of support should identify the problems facing the students in terms of issues of school population growth and lack of space. Study participants also thought letters concerning the educational philosophy of the school, the outdated nature of present facilities, and the contributions of previous generations to the schools were effective.

The opposition also sent letters to the editor in 75 percent of the communities attempting to pass a school bond referendum. In response, supporters focused on the cost/benefit ratio of the tax increase (e.g., image of the community and attracting new businesses).

Personal Campaigning

Survey participants from school districts that successfully passed bond referendums believed that one-on-one campaigning, such as door-to-door canvassing and telephone campaigning, was effective in achieving a positive result in their campaigns. They also identified public meetings as a worthwhile activity.

Variables That Contribute to the Failure of School Bond Referendums

The study also revealed variables identified by the campaign leaders as having a negative effect on election results.

Lack of Understanding

The most consistently identified variable contributing to a failed bond referendum was a lack of understanding of the attitudes and perceptions within the community and among educational staffs about the schools. When proponents of the proposal failed to identify negative attitudes and address them, opposition groups emerged. Most participants in the study from communities where bond referendums failed identified these groups as a primary reason the bond issue failed. The proponent groups could not convince the public that the present facilities were inadequate because the individuals did not realize that many citizens had a different view of the existing schools.

School Board Support

The study indicated that the school board decision to proceed with the referendum must be unanimous. A negative vote by a board member apparently sends a message to the public that something is wrong with the proposed project, and that it does not warrant support at the polls.

Size of Increase of Tax Levy

Study participants from school districts where referendums were defeated indicated that a primary reason for their loss was taxpayer concern over the raise in taxes. The data suggest that the amount of levy increase per one thousand dollars valuation must not exceed $2.00. School boards may need to consider the use of capital outlay dollars as a means of lowering the levy increases requested.

Influential Variables Specific to the Characteristics of a Given School District

Every school district has its own set of unique issues that may affect the outcome of bond elections. The study indicated that voters have a wide array of concerns.

Placement of School Building

In some school districts, the location of the proposed school building had an influence on passage of the bond issue. This was particularly true in consolidated school districts or in districts where relocation of the building meant longer bus rides for students attending school in a different area of the community.

School Design

The study participants in some districts indicated that it is a mistake for school boards to offer the simplest school building design on the first bond attempt. When this occurs the school board has nowhere to cut spending on the project for the second attempt.

Demographics

Most participants indicated that the composition of the population of the school district had a direct effect on passage or failure of a bond referendum. In many communities, less than 35 percent of the voting adults had school-age children. Thus, meeting the 60 percent approval level required was a formidable task. Most respondents denoted a need to gain the support of senior citizens.

Perceptions of the Economy

In some communities, citizens' perceptions of the economic future of their communities had a critical impact on election outcomes. Therefore, most connoted the need to supply information to the public about the potential growth of the community and the impact a new school might have on attracting people to the community.

RECOMMENDED ACTIVITIES FOR SUCCESSFUL BOND REFERENDUM CAMPAIGNS

The following recommendations are based on the review of relevant literature and on the results of the study discussed above. They provide information complementary to the step-by-step plan for passing bond issues that will be covered in the remainder of the book.

1. The superintendent should ensure a unanimous vote of support by the board of education. This may mean proponents of the project

must make some concessions, but failure of the board to reach a consensus appears to doom the results of the election.

2. Administrators and board members should keep as low a profile as possible. While these individuals must supply organization, information, and other support to the citizen support group, they must allow the "ordinary" person to carry the message.
3. The board and administrators should establish a diverse community task force. This group can supply valuable information about public perceptions and can be instrumental in providing appropriate information to the general public.
4. The attention of the campaigners should be on "yes" voters. Political campaigners learned a long time ago that those people who have decided to vote "no" on something do not change their minds easily. Often trying to turn a "no" vote into a "yes" vote is futile and the attempt can eat up resources better utilized elsewhere. Proponents should concentrate on getting "yes" voters to the polls and convincing the undecided to vote "yes."
5. The local media and school staff members should be involved in the early planning stages of the campaign.
6. School boards should utilize experts such as bond consultants, architects, and other trained individuals to educate support groups in the community.
7. The citizens committee should concentrate a great deal of effort on disseminating information. Flyers, brochures, question and answer sheets, and other printed materials are important tools for getting the message out. One should be cautioned to keep the materials simple and straightforward. Complicated, jargon-ridden justifications merely cloud the issue. Telephone and door-to-door campaigns are also important. Most people like to be able to ask specific questions, but many will not do so in a public meeting.
8. Collaborate with other governmental agencies. Sometimes bond issues are easier to sell to the public, and different state laws may apply if renovation and construction are combined with facilities operated by other governmental agencies. State laws vary on the issue.
9. The school board should limit the tax levy increase by keeping the school design simple and by utilizing existing capital outlay funds as much as possible.

10. The information disseminated and public relations activities should focus on the benefits to children and the community. Supporters need to explain the benefits of a quality education to the entire community, how better facilities contribute to providing a high quality education, and why residents whose children have left the community should care about today's youth. While a new facility might make life better for teachers and administrators, these benefits should be downplayed because they do not evoke public empathy.
11. School boards and administrators should seek advice from administrators and school boards that have won bond elections. No two school districts are alike; each faces its own unique problems, but it is important for administrators to contact other administrators who successfully organized bond referendum campaigns. They should find out what techniques and activities worked for others and then adapt those strategies to their own situations.

One-third of the public schools in the United States, housing about 14 million students, is presently inadequate to meet the educational, health and safety needs of those students. Research indicates that neglect of school building maintenance, unprepared school administrators, and an anti-tax mindset are factors that will continue to contribute to that inadequacy. To counteract the above barriers, school administrators and boards of education must acquaint themselves with sound strategies for passing bond issues for repair, renovation, and new building projects.

In actuality, school administrators, school board members, and school supporters must devise and implement a four-phase program. The first phase is the preplanning phase, during which the facility needs of a school district's educational program are assessed. The second is the project development phase, during which consultants and others develop plans for the building project. The third phase is the campaign phase, during which school supporters conduct a campaign to win approval of the project from the voters. The final phase is follow-up, during which the reasons for a failure are assessed or those activities associated with implementing the building plan are carried out.

The chapters that follow identify specific activities for each of the four phases in a school bond issue campaign. No suggestions for formulas can guarantee success; however, developing an organized plan of action can increase the probability of success when it counts the most.

REVIEW ACTIVITIES

1. Identify the facilities and structural needs of your school district.
2. Describe the factors that contribute to your school district's physical conditions.
3. List the factors that have served as barriers to rehabilitation, remodeling, renovation, renewal, and modernization in your district.
4. Find how your district has financed facilities in the past. Has your district opted for pay-as-you-go, reserve funds, bonding, or other methods? What contributed to these choices of financing in your district?
5. Compile a list of representative community groups in your district. For each group, identify motivating factors that would cause support for a bond issue campaign.
6. Consider possible expanded functions for your school district to meet needs of the community. How would each function increase positive public relations?
7. Identify influential people in your community, excluding political figures, and determine how each can increase the potential for school bond victory.
8. Specify ways to achieve school staff and board of education unity.
9. Project ways to increase voter turnout in your school district.
10. Analyze Condorcet's reliability formula (see Exhibit 1.1).
11. List the states that require two-thirds, four-sevenths, three-fifths, or one-half plus one of the votes cast to pass a school bond referendum.

ENDNOTES

1 United States General Accounting Office (GAO/HEHS-95-61). 1995. *School Facilities: Condition of America's Schools*. Washington, DC: United States General Accounting Office, p. 1.

2 GAO/HEHS-95-61, p. 1.

3 pp. 10–11.

4 ———. pp. 9–10.

5 ———. p. 10.

6 ———. p. 11.

7 ———. p. 11.

8 ———. p. 11.

9 ———. p. 11.

10 ———. p. 11.

11 ———. p. 11.

12 ———. p. 12.

13 ———. p. 12.

14 ———. p. 12.

15 Research report in Ortiz, F. I. 1994. *Schoolhousing: Planning and Designing Educational Facilities*. Albany, NY: State University of New York, p. 32.

16 ———. p. 32.

17 "District Sued Over Classroom IAQ." 1997. *American School & University* 69(12): 3.

18 United States General Accounting Office (GAO/HEHS-95-61). 1995. *School Facilities: Condition of America's Schools*. Washington, DC: United States General Accounting Office, pp. 35–36.

19 Smith, L. 1997. "School Averts Tragedy When Chimney Tumbles." *Argus Leader*, 18 February 1997, A1.

20 ——— . 1997, A1.

21 United States General Accounting Office (GAO/HEHS-95-61). 1995. *School Facilities: Condition of America's Schools*. Washington, DC: United States General Accounting Office, p. 11.

22 GAO/HEHS-95-61, p. 17.

23 ———. p. 17.

24 ———. pp. 5–6.

25 ———. pp. 6–7.

26 ———. pp. 6-7.

27 Kosters, H. G. 1997. "S.D. School Facilities Report." *Bulletin*, 49(5): 1–2. [Study commissioned in 1995 by the South Dakota Department of Education and Cultural Affairs and the Associated School Boards of South Dakota. Summary of the study was presented to the South Dakota Legislature during the 1995 legislative session.]

28 ———. 1997.

29 Agron, J. 1997. "Rising to New Heights," *American School & University*, 69(9): 22.

30 Gerald, D. E. and W. J. Hussar. 1994. *Projections of Education Statistics to 2005*. Washington, DC: U.S. Department of Education, National Center for Education Statistics.

31 Burrup, P.E., V. Brimley, Jr., and R.R. Garfield. 1993. *Financing Education in a Climate of Change*. Fifth edition. Boston: Allyn & Bacon, p. 292.

32 ———. p. 292.

33 ———. p. 293.

34 Alexander, K. and R. G. Salmon. 1995. *Public School Finance.* Boston: Allyn & Bacon, p. 337.

35 Burrup, P.E., V. Brimley, Jr., and R.R. Garfield. 1993. *Financing Education in a Climate of Change*. Fifth edition. Boston: Allyn & Bacon, p. 294.

36 South Dakota Codified Laws Ann. 13-16-2 (1991). Types of funds enumerated. All school district funds shall be placed in either the general fund, capital outlay fund, special education fund, public service enterprise fund, trust or agency fund as defined in 4-4-4, bond redemption fund or 874 fund as hereinafter defined.

37 Colorado State Statute 22-42-120. Internet Address: http:ww.ppld.org/CoStatutes/1220/1220042001200/html.

38 McLean, I. and F. Hewitt, eds. 1994. *Condorcet: Foundations of Social Choice and Political Theory*. Brookfield, VT: Edward Elgar Publishing Company.

39 Holt, C. R. 1993. "Factors Affecting the Outcomes of School Bond Elections in South Dakota," Ed.D. diss., University of South Dakota, pp. 9–50.

40 ———. pp. 9–50.

41 ———. pp. 9–50.

CHAPTER 2

The Preplanning Phase

THE key to dealing with any problem or project is effective planning. A checklist of preparations that should be completed throughout the campaign and election processes is included in Appendix B. When dealing with providing appropriate facilities in which to implement a school's educational program, school administrators and board members will deal with both ongoing and project-specific activities. An exposition of "Planning in Debt Issuance," as presented in Appendix C, depicts the importance of planning and controlling a school district's operation by a school board. While this book focuses on development and campaign stages of a specific project, it is important to discuss some of the key elements of the preplanning phase and the impact they have on the project as a whole.

THE SCHOOL FACILITIES PLAN

Any building project within a school district should be viewed as merely one part of an overall educational plan. Therefore, the school superintendent and school board should have a clear picture of what they want to offer within their educational program and should develop a facilities plan based on how school buildings can help educators meet the goals of that program. For example, new teaching methods brought about by the introduction of computers and other electronic equipment require different sizes and shapes of classrooms and other facilities (a good library now includes several student carrels with computers); mainstreaming disabled students necessitates careful thought as to classroom size, hallway placement, and bathroom arrangements; and new safety requirements require special ventilation and toxic substance control equipment in science labs.[1] New educational methods (e.g., open classrooms, lab inquiry methodologies, and integrated units developed by teams) have spatial and arrangement requirements that may differ significantly from the traditional classroom. For example, if a school is utilizing one of the laboratory inquiry methodologies for teaching science, the amount of laboratory space required for each student increases by about 25 percent over the traditional laboratory.[2] Facilities planning, then, is not just a matter of deciding which buildings are unsafe or unhealthy; it is a process of identifying which buildings no longer serve the educational mission of the school district and how best to modify or replace the structures.

Of course, no one would recommend scrapping a structurally sound, 30-year-old building simply because the classrooms are not the right shape for certain teaching methods. But such a consideration might be a legitimate justification for remodeling or renovating. The key is to understand what is needed and to plan now for future projects.

Most school superintendents find it useful to establish a five-year building plan. The plan should include potential new building projects, major renovation and repair projects, minor repair projects, buildings that might be abandoned, and potential uses for buildings no longer utilized for classes. Such plans should also include any shifts in attendance centers (i.e., if a new high school is built, can the overcrowded middle school move into the old high school). This plan should be reviewed on an annual basis, and all persons involved should understand that the board may make revisions based on current circumstances.

To be successful, school administrators cannot do such planning in a vacuum. School officials must constantly seek input from administrators, teachers, parents, and members of the community as to their needs, the problems they see in implementing programs in present facilities, expectations they have of their school's educational programs, and their ideas about the relationship between the school and the community.

To gain such input, a school administrator must develop two programs:

1. An ongoing community relations program that encourages parents and others to voice their concerns about the school system and to bring ideas to top administrators

2. An internal system through which other supervisors, teachers, and support staff personnel can address their needs in terms of facilities and present their ideas about solutions

School administrators should design these systems to allow both faculty and laypersons to understand the issues facing the school districts (e.g., budget limitations, the impact of state and federal mandates, changing curricular needs, changes in school population demographics) and express their views as to the priorities of the school. For example, if parents and faculty believe the students' learning is being affected seriously by violence in the schools, they probably would name a security system as their top priority rather than installing fiber optics. A board of education that avoids the first concern of the parents and teachers because it "sounds negative" is unlikely to gain community support for other measures. The lesson here is "to listen, and then demonstrate you heard."[3]

THE COMMUNITY RELATIONS PROGRAM

Most school administrators find that an ongoing community relations plan is essential to the eventual passage of bond referendums. This type of program can provide invaluable information about public perceptions of the school's needs, the attitudes of individuals who might oppose bond issues, and the enthusiasm of individuals who would likely support a bond issue. While some school administrators choose to obtain such information through informal coffee klatches or visits with influential members of the community, other superintendents find regular "town meetings" more effective. Special invitations to parents, city council members, well known citizens, and senior citizens encourage the participation of a cross-section of the community.

School administrators also should use the news media to present a positive image of the school and its programs and to identify and clarify specific educational concerns. While most newspapers and electronic media cover special events and sports, many newspaper editors need to be encouraged to print stories about unusual programs being developed in the schools or problems confronting the educational establishment. For example, a special report on the difficulties of bringing needed computer technology into every classroom assists the general public in understanding the needs of a modern educational system. School superintendents may want to write a regular column for the local newspaper in which they discuss upcoming issues—both achievements and problems.

Some administrators fear exposing problems to the public; however, one must bear in mind that if the public is not aware that a problem exists, they will not understand why the school board is "suddenly" asking for more money to solve the problem. A building project that appears "out of the blue" is unlikely to gain sufficient support for passage.

Any good community relations program emphasizes the school's role in the economic development of the community, the importance of a quality educational system to all members of the community, and special services the school provides to various segments of the community. For example, if the community is trying to attract new businesses, school administrators should inform the public about the high priority many corporations place on the quality of the local educational system when selecting a site for developing their businesses. Other school systems may want to point to community education or GED programs that benefit individuals from all age groups within the community: their point being that schools are not just for children anymore.

In some school districts, the school board publishes an official newsletter once a month and distributes it to their students to take home to their parents. They also make the publications available at various stores and service offices in the community. This endeavor can be somewhat expensive if commercially done; however, using new computer publishing programs to generate the documents in-house can reduce the price considerably. Some school superintendents indicate that this newsletter may have been a primary factor in winning a school bond election. The newsletter should contain stories about issues confronting the school board, explanations of decisions made by the school board, positive stories of accomplishments by either students or faculty, and announcements of important future events.

DEVELOPING AN INTERNAL SYSTEM FOR INPUT

School administrators also know that those persons closest to a problem are usually the individuals who know the most about it. Therefore, supervisors seeking information about the adequacy of school facilities are served best by asking those individuals who must work in the facility on a day-to-day basis: the teachers and support staff. Administrators can employ several systems to gain the appropriate input. Some superintendents find it most useful to hold special meetings with building and department supervisors for the purpose of discussing specific needs. Other top school officials find they receive better information by meeting directly with small groups of faculty and staff members or with individuals.

From these sessions, the school superintendent should gain an understanding of concerns about health or safety, the logistics of implementing new programs in present facilities, overcrowding problems, introducing new technologies, and a lack of space for special services or projects. Some school administrators also find it useful to include board of education members in these special needs sessions, so they also are aware of conditions within the schools.

DESIGNING AND UTILIZING A SCHOOL FACILITIES SURVEY

One other method for obtaining input from both educational staff and laypersons in the community is the development of a school facilities survey. The project can be used to: (1) determine what school buildings the district needs; (2) determine how present facilities can be utilized best; and (3) suggest a school building program that provides for additional facilities.

School facilities surveys can be conducted by local groups, the school superintendent and board of education, professional staff persons, or parent/teacher organizations. The board of education might engage a survey team composed of educational consultants to interpret the results of the surveys and describe the best course of action. Most state departments of education employ persons who can assist school districts in determining their needs.

The ultimate goal of the survey should be to identify the concerns, needs, and expectations of educational staffs, parents, and other citizens in terms of educational programs and health and safety needs. Areas of concern that could be part of the survey include:

1. What does the community want its schools to provide in terms of educational programs? What types of academic subjects should be

taught? Should there be a grade school music program? Art? Should the schools provide night classes for adults? What new technologies should be included in the school environment? Are there special conditions within the community that would warrant teaching specific types of courses (e.g., teaching agriculture classes in a rural area)? What extra-curricular programs are important to the community?

2. What are the present space needs of the educational program and how will those needs change in the future? What is a reasonable number of students per classroom? Are there unusual needs within the community that require special considerations by the school district (e.g., a sheltered area for students to gather before or after school hours because most parents in the community work)? What demographic changes are taking place in the community that might affect enrollment? Are the present schools' playgrounds large enough to permit organized and supervised activities?
3. Is the existing school plant safe and cost efficient to operate, and is it suitable for integrating new technologies and curriculum into the educational program? What is the building's age? How is the building presently wired? Have roofs, chimneys, and other structural features been inspected recently? Does it take an exorbitant amount of fuel to heat the building in winter? Are there features that do not meet current health and safety codes? Are the classrooms, bathrooms, and other necessary facilities accessible to students with special needs?

Prepared by: ______________________
Position: ______________________
City: ______________________
Date: ______________________

BUILDING: ______________________

Room No.	Location (basement, 1st, 2nd)	Area (square feet)	Usage (grade or subject)	No. of pupil stations (desks, etc.)	Maximum No. of pupil stations room can accommodate (divide area by 30)	Over-flow

Adapted from the Survey Division of the Bureau of Educational Research, Ohio State University.

Figure 2.1. *Sample elementary school plant data sheet.*

Prepared by: ______________________
Position: ______________________
City: ______________________
Date: ______________________

ROOM AND PUPIL-STATION UTILIZATION OF SECONDARY SCHOOL BUILDINGS

BUILDING: ______________________ ROOM NUMBER: ____________

1. Location of room: __________ floor (basement, 1st, 2nd, 3rd, 4th, etc.)

2. Dimensions of room: __________ feet by __________ feet = __________ square feet

3. Number of pupil stations now in room ______. (A "pupil station" is a desk, chair, or working space for one student.)

4. How many pupil stations, if any, could be added to this room?__________

5. How many pupil stations, if any, should be removed from this room?__________

6. Using the chart below, provide information relevant to the daily use of this room. In the first column, indicate the number of minutes for each separate usage. In the columns headed by days of the week, indicate the subject taught during that time period or a special use of that room and the number of students in the room during that class or activity. For example, if the room is used during the first period every day of the week for Freshman English and there are 22 students in the class, write English 1--22 in the columns for the days of the week (If it is used for the same activity every day, you may simply draw an arrow from Monday through Friday). Do not report homeroom periods of less than 30 minutes or after-school use of the room. If the room is not in use during a certain time period of a regular day, write the word "vacant" in that slot.

Period	Length	Subject/Activity Number of Students				
		Monday	Tuesday	Wednesday	Thursday	Friday
1						
2						
3						
4						
5						
6						
7						
8						
Other						

Adapted from the Survey Division of the Bureau of Educational Research, Ohio State University.

Figure 2.2. *Sample secondary school plant data sheet.*

Various authors suggest different plans for conducting building surveys. Figures 2.1 and 2.2 provide examples of school plant data sheets that elementary and secondary staff members can fill out to help school officials assess the condition of their present classroom and work areas. These surveys also can provide information as to which classrooms may need to be inspected immediately, which schools seem to be the least adequate in terms of prioritizing, and which school buildings are meeting the needs of staff and students. Regardless of the type of survey instrument the school district utilizes, the focus should be on determining how well various school buildings meet the needs of the educational program.

DESIGNING A SCHOOL FACILITIES PLAN

The second part of adequate planning for a school district involves establishing a facilities development plan. The first step of such a plan must be establishing board policies as to facilities development goals and facilities planning. Such policies (see examples in Figures 2.3 and 2.4) should reflect the priorities under which the board will make specific decisions and the process by which the board will make decisions about repair, renovation, and new construction projects.

After adopting appropriate policies, school administrators and boards should develop both short- and long-term plans for capital development in their districts. The school board should base such plans on input from supervisors, teachers, other staff, and laypersons brought about through meetings and/or the plant-need survey. The school superintendent also should make recommendations based on the life expectancy of facilities, enrollment projections, changing instructional

The Board accepts the premise that a school building should reflect the philosophical convictions of the school district about education. The Board recognizes, however, that educational programs are neither unchanging or simple to incorporate into a facility plan. Anticipation of program change makes the need for flexible use of buildings necessary.

School buildings will be functionally compatible with desired school experiences. The program, not the physical setting, will dictate the manner in which the building is used.

The Board recognizes that funds are limited, and that when planning facilities priorities must be established to make the best use of the school building dollar. The Board's first objective will be to develop a plan that provides adequate space for each student's educational development. Whenever possible, the cultural as well as educational needs of the community will be considered in planning facility expansion.

Architects retained by the Board will be expected to plan for simplicity of design; sound economics, including low long-range maintenance costs, efficiency in energy needs, low insurance rates; high educational use; and flexibility.

(Adoption date, May 10, 1982)

Source: Brandon Valley School District 49-2, Brandon, SD. Used by permission.

Figure 2.3. *Sample—facilities development goals.*

The Board is responsible for the regular operation and orderly development of its physical plant. For this reason, the Board will concern itself with both short- and long-range planning as it relates to the properties of the school district.

To this end, the Board will follow the policy of having before it at all times a long-term building program to serve as a guide for capital improvements. In developing a long-range program, the Board will monitor:

1. The evaluation of existing facilities in terms of capacity and function.

2. The projection of life expectancy of facilities and maintenance costs.

3. Enrollment projections and community development patterns.

4. Site availability and acquisition.

5. Changing instructional requirements and services.

This program will be subject to systematic study, revision, and extension from time to time, and the respective construction projects will be acted upon individually when proposed for implementation.

The Board's building program will be designed to provide adequate facilities to conduct full-time elementary and secondary education programs for all students residing in the district. The building program will be based upon specific Board policies and have been and will continue to be modified to conform to changes in the curriculum, availability of construction funds, and changes in enrollments.

(Adoption date--May 10, 1982)

Legal Ref.: SDCL 13-24-9

Source: Brandon Valley School District 49-2, Brandon, SD. Used by permission.

Figure 2.4. *Sample board policy—facilities planning.*

requirements, and new or current health and safety requirements. The superintendent also makes projections as to the availability of funds and estimated costs of projects so that the board is able to utilize the most suitable form of funding. The plans should be reviewed and updated yearly.

When making decisions about both long- and short-term projects, those involved must consider various factors. Uppermost among these is the critical need factor.[4] Projects that would bring school buildings into compliance with health and safety codes (e.g., lead paint removal) probably would be of higher priority than projects that deal with conveniences (e.g., enlarging a parking lot). Renovation and repair projects for buildings that have received little attention in the recent past would have priority over projects in schools that had received capital outlay funding recently. School administrators and board members also need to factor in the need for support of future expenditures, the value of the project to the overall educational program, and the relative costs/benefit ratio (e.g., Is it really less expensive to renovate an aging building than it is to build a new one? see Chapter 5).

Both the superintendent and the board face many dilemmas when developing a facilities plan. For example, the teachers in a particular elementary building believe that a top priority for their school would be wiring for Internet access, while the parents from another building want the district to repave the playground area. The school district can do only one of the projects this year. The board must determine which group must wait. On the one hand, Internet access would open new curriculum areas and would provide a much-needed resource for information; on the other hand, some parents have been complaining about scrapes and bruises caused by the rough surface of the present playground. The school that wants Internet access is in a relatively affluent neighborhood and the school received a major renovation only three years ago. The other school is located in a mostly lower-middle-class neighborhood with a large minority population; the facility has not received capital outlay funds for more than seven years and is not slated for any renovations for at least three more years. Neither project is an emergency. Even though no major issue is involved, the example shows how difficult such a decision might be in light of the various factors that must be considered.

It is vital to any school district that the school board and administration make the details of both the long- and short-term facilities development plans available to the public. Additionally, the public needs to know the justifications for expenditures, the projected costs of projects, and the method the board will utilize to pay for the projects. This is especially true if a major capital-outlay project is planned that might require a bond issue.

Careful planning has always been the hallmark of successful projects, and it is no different for bond issue election campaigns. School administrators and boards of education must lay the groundwork for gaining public acceptance for capital outlay projects early on if they are to expect positive results.

MAKING DECISIONS ABOUT SPECIFIC BUILDING PROJECTS

While the above steps in the process can go a long way toward building community support for capital outlay projects in general, school administrators and boards need to take more specific steps that will help them make decisions about proceeding with a specific building project, especially if it will require a bond election.

Probably the most important information the school superintendent can gather at this time concerns alternatives for solving the problem. The answer to every facility problem is not necessarily building a new school.

Some schools need to be rehabilitated. The building is brought up to the condition it would be in if all maintenance had been done. In this case, the school district is doing all the maintenance at the same time. This solution is viable for structures that are fairly new (not more than 20 years old) and which remain structurally sound.[5]

In some instances, the facilities can be remodeled. Remodeling involves changing some internal and external spaces and/or changing the location of certain functions within the building. This alternative often is used when the structure is stable and large enough, but the educational program requires a different utilization of space.[6]

Sometimes the usefulness of a building is questioned to the extent that the only alternative to building a new structure is renovation. In these cases, the utilities, structures, and design of the building are changed to the extent that they can meet the demands of a modern school building. This is the most complex of the alternatives, and the most expensive.

Many school boards look to renovation or renewal because, on the surface, these options appear to be less expensive than building a new structure and time to occupancy seems shorter. Neither of the above may be true. School planning experts have developed several methods of analyzing which approach, renovation or building, would be the recommended solution. Some recommend comparing the total cost of the project, indicating that if the estimated total cost of renovation is more than 50 percent of the cost of building a new structure, renovation is probably not the best alternative.[7] Others, however, indicate that a lower figure (40%) may be a better level at which to evaluate the two.[8] Castoldi has developed the Generalized Formula for school modernization.[9] This formula takes into consideration total cost of educational improvements, total cost for improvements in healthfulness, total cost for improvements in safety, estimated index of educational adequacy, the estimated useful life of the modernized school, cost of replacement of school considered for modernization, and estimated life of the new building.

Earthman[10] suggests that most mathematical formulas do a fine job of comparing quantifiable data, but they fall short when non-quantifiable information is needed to complete the formula. In other words, how does one actually put a number to a judgment such as educational adequacy? Earthman suggests that a four-part analysis, which includes measuring the adaptability and condition of the present facility, measuring the adequacy of the present school site, analyzing the cost/benefit ratio of renovation, and determining the emotional issues that are tied to the building. If renovation is a consideration, the ability to choose that recommendation must be available to the building committee (see Chapter 5).

The school superintendent also should be gathering as much information as possible about community attitudes toward the need for the specific project—be it building a new structure or renovating an existing structure. These data will inform the superintendent of factors he or she and the board of education need to consider before proceeding. First, the data will reflect citizens' perceptions and concerns about economic conditions in the community. For example, if a community has just experienced a major factory shutdown or layoff, everyone in the community may be experiencing anxiety over their economic futures. People who do not know whether they will be able to make their mortgage payments are not likely to vote to increase their property taxes. If there is a negative economic environment, it probably is best not to hold a bond election.

Likewise, if most people in the community believe the school is being managed well, there is a likely chance many individuals will support a building project. If, however, the school superintendent continually hears complaints about discipline in the schools, expensive projects that benefit only a few students, administrative salaries, or wasteful spending practices, he or she may find that trust levels are not high enough for many people to "risk" their money.

The school superintendent also should note divisions within the community. Racial or class strife within a community can bring devastating results at bond election time, particularly if the building project might change the demographics of the schools (i.e., a school that is now predominantly of one ethnic group changes to one in which most students are from a different ethnic background). While persons of goodwill believe that no one should have to consider this factor in a democratic society, such is not the reality.

School enrollments are another complicating factor, and the administration needs to gather data on the potential student numbers in various parts of the community. School officials can use several methods for estimating potential student populations. It is beyond the scope of this book to provide an analysis of these processes. Suffice it to say that it is not an easy project.[11] The superintendent needs to gather information as to potential population growth or decrease in his or her community, how changes in demographics in the district will affect the number of students enrolled in the schools, potential economic growth or decline, and fertility rates. While natural or man-made disasters that could affect student populations cannot be anticipated, one can arrive at a fairly close estimate of future student enrollments by using one of several formulas recommended by educational statisticians. Computer software is also available.

While no one factor should hold up pursuing a building project, a prudent person will take all factors into consideration before proceeding. The development of building projects and preparing for a bond election are expensive processes. If certain conditions exist that will predetermine a negative result, the school board might be wiser to spend that money on repairs or small renovation projects.

The superintendent also should be gathering input from board of education members about their attitudes toward the building project. One of the critical variables to bond referendum passage mentioned in Chapter 1 is a unanimous vote of endorsement for the bonding by members of the board of education. Before the superintendent proceeds further in the process, he or she should be confident that all board members will agree that they should proceed to the preliminary stages of preparing for a bond election.

The administration might choose to poll area legislative representatives and county and city officials to determine their attitudes toward the school district's needs. Such individuals might be willing to share information about citizens' concerns.

If the school superintendent and board of education determine that the timing is as good as it is going to get in terms of developing a building project and conducting an election, they should begin the formal process by addressing the issue at a public board meeting. In keeping with the findings of the studies cited in Chapter 1, the superintendent and board members should maintain as low a profile as possible, even at this stage of the project. They may find it most beneficial for a citizen to bring the concern to the board at a public meeting. The superintendent probably will be able to identify an appropriate person for the job. The individual needs to be respected by many segments of community and be willing to go before the public and media to present the idea.

At this board meeting the superintendent needs to be prepared to discuss some basic justifications for the project, input he or she has already obtained about a building project, and a plan for obtaining more input from a cross section of the community. The board should authorize project development activities. This phase can include gaining more specific input from educational staff, community leaders, parents, and others, developing a financial report that reflects the amount the district can afford, and developing preliminary plans for the building in terms of educational specification. The board also may choose to have the superintendent appoint committees to pursue various aspects of the development.

The school administration and school board should begin involving the media at this point. Most media outlets will conduct interviews and prepare stories on such proposals and present them in a fairly objective manner. After newspaper, television, and radio reports first appear, the opposition usually emerges. Letters to the editor opposing the proposal usually are printed one or two days after the proposal has been announced. From the letters, administrators can begin assessing the major reasons for opposition and from which groups most of the resistance to the plan will emerge. For example, if a prominent individual from a minority segment of the population comes out in opposition to the referendum, one probably can assume opposition from that segment of the community. One cannot assume, however, that opposition is based solely on property tax increases. It may be, in fact, that some parents believe the school their children attend is badly run down. They might believe their interests are being ignored in favor of those of another segment of the community. One should never assume to know the reason for opposition without asking.

The school board also may find a community action committee helpful at this time. School administrators and the community lead person should identify persons within the community who would likely support the bond issue. These individuals could be asked to serve on a community action committee. This group's responsibilities would include obtaining information about the potential number of "yes" voters there might be in the community, the concerns voters have about a bond issue at this time, and individuals who might be willing to serve on committees during the development stages. The process should not be formal. Generally, those on the action committee form a network through which they discuss the proposed project one-on-one with citizens in the community. The systems can include forming telephone trees, interviewing civic leaders, or talking to acquaintances at meetings or other gatherings.

The most important aspect of these activities is determining whether there is sufficient support at this time to continue with the process. If fewer than 45 percent of those contacted give positive or moderately positive responses, the chances are small that the super majorities required in most states for passage can be reached, even with an extensive education campaign. While a 45 percent figure might appear high, one must remember that, in these situations, "yes" votes are usually over reported.[12]

While this type of surveying cannot be considered scientific polling, the type of questions to be asked should be planned ahead of time and the volunteers should be trained as to how to ask the questions and record answers. If the polling data is not objective, the results obtained are actually useless in the decision making process. Information about writing and conducting surveys can be found in Chapter 8 of this book. Also, to obtain objective information, volunteers should be assigned to all parts of the community so that a true cross section of the population is covered.

Volunteers should be trained not to take time to discuss the issue with any person who answers "no" to the question of supporting the bond issue and they should be trained in how to effectively deal with hostile individuals (see Chapter 8). When volunteers receive a positive response, they should ask the person whether they would be interested in talking to some of their friends about the proposed project or if they would be willing to help fund the formal campaign. Volunteers also should write down the names of individuals who indicate they will vote "yes," as these persons are those from whom volunteer efforts can be solicited during the campaign.

A separate team of individuals should tabulate and interpret the information. The most useful categories of analysis would include: percent of "yes" voters, number of potential "yes" votes for selected age and ethnic groups, areas of the school district with either high or low percentages of "yes" voters, and special considerations unique to the district (e.g., rural vs. city).

The interpretation team should present the information to the superintendent and board members. If the information is positive, the board should authorize proceeding to the project development phase.

REVIEW ACTIVITIES

1. Research the building and maintenance history of the buildings in your school district. Also, do a detailed listing of how your buildings do not meet standards for compliance with state and federal mandates and reasons buildings are outdated for present curricular needs.
2. From the research on building and maintenance history and personal inspection, develop a list of problems that will need to be attended to in the next five years.
3. Establish a priority ranked list of capital outlay projects for the next five years.
4. Estimate how your school district could financially deal with the above capital outlay projects.
5. List conditions in your community (economics, anti-tax groups, and so on) that would interfere with your school district's ability to pass a bond issue.
6. List those persons or organizations within your community from which you could likely obtain support or volunteers for preliminary surveying.
7. Evaluate the preparations that should be completed by election day depicted in Appendix B.
8. Summarize the planning and controlling process in debt issuance illustrated in Appendix C.

ENDNOTES

1 Castoldi, B. 1994. *Educational Facilities: Plan, Modernization, and Management.* Fourth edition. Boston: Allyn & Bacon, pp. 71–74

2 ———. pp. 71–74.

3 Carter, M. A. 1995. "How to Blow a Bond Issue—Or Not, If You'd Prefer." *Clearing House*, 68(5): 289–292.

4 ———. pp. 289–292.

5 Earthman, G. I. 1994. "Scrap It or Rehab It: A Process for Deciding When to Renovate." *School Business Affairs*, 60(1): 4–8.

6 ———. pp. 4–8.

7 ———. pp. 4–8.

8 Castoldi, B. 1994. *Educational Facilities: Plan, Modernization, and Management.* Fourth edition. Boston: Allyn & Bacon, pp. 71–74.

9 ———. pp. 71–74.

10 Earthman, G. I. 1994. "Scrap It or Rehab It: A Process for Deciding When to Renovate." *School Business Affairs*, 60(1): 4–8.

11 Castoldi, B. 1994. *Educational Facilities: Plan, Modernization, and Management.* Fourth edition. Boston: Allyn & Bacon, pp. 71–74.

12 Holt, C. R. 1993. "Factors Affecting the Outcomes of School Bond Elections in South Dakota," Ed.D. diss., University of South Dakota, pp. 9–50.

CHAPTER 3

The Project Development Phase: Selecting Consultants

AFTER the board of education has voted to proceed to a bond referendum election, they will need to seek the assistance of both a bond consultant and an architect. Using these experts will save both the school superintendent and members of the school board much anxiety and will give patrons of the school district the security of knowing the facts and figures they are given are accurate.

SELECTING THE BOND CONSULTANT

The Bonding Process

The issuance of bonds is a process regulated by the state governments. In order to understand the importance of hiring a bond consultant, it is important to understand the bonding process and those elements that might contribute to problems. To assist with understanding the bonding terminology see Appendix D, Glossary of Terms: Municipal Bonds.

General Obligation Bonds

General Obligation Bonds, also referred to as G.O. Bonds, are bonds secured by the issuer's pledge of its full faith, credit, and taxing power for the payment of the bond. School districts have the power to borrow money on the credit of the school district for any authorized corporate purpose within the constitutional limitations of municipal indebtedness. In most states, school districts are authorized and empowered to issue negotiable bonds for the following purposes only:

1. To refund any bonded indebtedness which is or is about to become due and payable, or whenever such indebtedness can be refunded at a lower rate of interest
2. To fund any judgments or outstanding warrants
3. To raise money for any purpose for which the school board is authorized to spend school district funds

A G.O. bond debt is debt for constitutional debt purposes only. In some states the debt of any school district can never exceed a designated percent (e.g., 10 percent in South Dakota) upon the assessed valuation of the taxable property for the year preceding that in which said indebtedness is incurred.

Capital Outlay Certificates

Capital outlay certificates can also be used to fund building. The capital outlay fund of a school district is a fund provided by law to meet expenditures of a designated number of dollars which results in the acquisition or lease of or additions to real property, plant, or equipment. Such expenditures are usually intended for the purchase of land, improvement of grounds or existing facilities, construction of facilities, additions to facilities, remodeling of facilities, or for the purchase or lease of equipment. In some states it may also be used for installment or lease-purchase payments for the purchase of real property, plant or equipment, which have a contracted terminal date not exceeding twenty years from the date of installment contract or lease-purchase, and for the payment of the principal of and interest on capital outlay certificates issued. The total accumulated unpaid principal balances of such installment contracts and lease-purchase and the outstanding principal amount of such capital outlay certificates cannot exceed a designated percent (e.g., 3 percent in South Dakota) of the taxable valuation. School districts have to provide a sufficient levy each year under the provisions of the statutes to meet the annual installment contract, lease purchase, and capital outlay certificate payments, including interest. The school board of any school district, where permissible, may issue capital outlay certificates to acquire or construct real property, plant, or equipment. All capital outlay certificates must be authorized, issued, and sold in accordance with the provisions set forth by state statute. In most states, unless specified otherwise, no election other than as provided by statute may be held, and the certificates may not have a maturity date in excess of the years specified (e.g., 20 years in South Dakota) from the date of issuance.

Other Types of Financing

Notes

Notes are promises to pay. In most states, any school board may borrow money, from any source willing to lend the money, by issuing a promissory note subject to the limitation and regulations set forth by statute. All notes must be authorized, issued, and sold in accordance with the provisions set forth in the state statute. Too, no election is held and the notes may not be issued for a term in excess of specified months by statute (e.g., 24 consecutive months in South Dakota). The notes may not exceed the sum of a designated percentage (e.g., 95 percent in South Dakota) of the amount of taxes levied or proposed to be levied by the school board but not collected at the date of borrowing by the school board pursuant to the state statute for the current or next full school fiscal year for the fund for which the money is borrowed, plus other anticipated receipts for the fund which have not been collected at the date of borrowing. In most states, notes pledging special education funds cannot exceed 85 percent of those expected funds.

Anticipation Notes

Anticipation notes are notes issued in anticipation of issuing bonds. Anticipation notes may mature no later than the specified years by state statute (e.g., 3 years in South Dakota) after the date of issue. Bond anticipation notes may be issued in a principal amount not exceeding the authorized principal amount of the bonds in anticipation of which the notes are to be issued. Such notes shall mature no later than a specified number of years (e.g., 3 years in South Dakota) after their date of issue.

Warrants

Warrants are promises to pay when sufficient money is available in the fund. Warrants are used when the school district has no money to pay debts.

Lease Financing

Lease financing is a means to finance a particular project or piece of equipment. A school district generally leases the property for a term and at the end of the term the ownership of the property transfers to the school district. Depending on the structure of the lease, leases are not considered debt for constitutional purposes.

GENERAL PROCEDURE IN ISSUANCE OF BONDS

The issuance of all school district debt is controlled by state law. State law determines what a school district can and cannot do. An attorney familiar with school district finance law should overview the issuance to make sure that all legal requirements have been met. Typically, nationally recognized bond counsel is retained to guide school districts through the issuance procedures and issue an opinion. Many issuers define a nationally recognized bond counsel as one who has issued a sole bond opinion in the last year.

When a school district is interested in issuing debt, a number of questions should be asked. For example, a school board should use the questions listed below as a checklist.[1]

1. Is the proposed project defined so that intended financing can be contemplated?
2. Are state constitutional limitations on debt financing applicable to the proposed bond issue?
3. Has the issuing entity been delegated specific authority by the state for the type of financing contemplated?
4. Does the proposed borrowing serve a proper public purpose?
5. Has proper approval been given by issuing entity through enactment of a resolution by its governing body?
6. Have special notice and public hearing requirements, if any, been observed?
7. Has the proper approval been secured from a state regulatory agency, if required?
8. Has an appropriate levy of taxes to pay the principal and interest been enacted, if such levy is required?

9. Has an appropriate pledge of revenues to pay the principal and interest been pledged, if such a pledge is required?
10. Have rates, fees, charges been pledged and are they sufficient to pay the principal and interest on the proposed issuance, if such is required?
11. Are any recitals and covenants in the bonds or notes made by an officer with the power to make them and are they adequate to protect purchasers?
12. Are the bonds or notes marketable and can they be sold at the proposed rates, terms, and amounts?
13. Have the bonds or notes been signed and delivered by the proper officers?
14. Have agents for paying, registration, execution, and other activities been properly appointed?

Need for an Approving Opinion

Generally, bonds are not marketable without an accompanying opinion of a nationally recognized bond counsel.[2] This opinion covers the following topics:

- whether the bonds are valid and binding obligations of the issuer
- the sources of payment or security for the bonds
- whether, and to what extent, interest on the bonds is exempt from federal income taxes and from taxes, if any, imposed by the state of issue

Role of Bond Counsel

The National Association of Bond Lawyers defines the role of bond counsel as follows:

> Bond counsel's role is one of an independent expert who provides an objective legal opinion concerning the issuance and sale of bonds. Bond counsel are specialized attorneys who have developed necessary expertise in a broad range of practice areas (e.g., school district, election, and bankruptcy law). Bond counsel often supplements rather than supplants the issuer's general counsel in protecting the issuer's interests. Irrespective of whether bond counsel represents the issuer or other parties in bond financing, bond counsel's opinion must be objective.[3]

Most bond issues require bond counsel because the buyers require that the bonds have an opinion of bond counsel on them. Governmental borrowings, all borrowings by Underwriters, Investment Bankers, and Bond Houses require bond opinions.

The Buyer of the Bonds

More important than the issuance is the sale of the bonds. The buyer of the bonds can be a private or public entity. Public is used in the sense that the entity is a subdivision of state or federal government.

Private buyers tend to be individuals, bond houses, investment bankers, banks, underwriters, and the like. Public buyers tend to be governmental authorities, agencies, pools, and departments. Common public buyers are the Health and Education Authority.

School districts can have a valid issuance of debt, but if no one buys the debt, it is of no benefit. Therefore, one of the main concerns a school district has is finding potential buyers for the bonds. Some of the more common buyers (South Dakota, for example) are as follows:

Underwriters
A. G. Edwards
Dain Bosworth, Inc.
Dougherty Summit Securities LLC
Kirkpatrick Pettis
Lehman Brothers
Piper, Jaffrey & Hopwood, Inc.

Financial Institutions
Banks

Governmental Units
Health and Education Authority
Rural Development
State of South Dakota
United States of America

Payment Terms

The terms on the bonds are important. They often determine whether the bonds are sold or not. Most often the terms are negotiated with the buyer; however, school districts also have the ability to put the bonds out for bid and have interested parties bid on the purchase. Other times the terms are preset, as in the case of governmental borrowings.

There are few limits on the terms of the bonds. The interest rate, maturity dates, payment dates, denominations, registration, medium of payment, and place of payment are whatever the governing body determines. However, the term is limited to fifty (50) years for bonds, two (2) years for notes, and three (3) years for anticipation notes. The school district must determine what terms are most beneficial.

Elections

Not all capital outlay expenditures are subject to election, but most that require bonds to be issued will have to be referred to the voters of the district. In other chapters, the authors cover the legal requirements and laws that control such elections. If the election is successful, the bonds can be issued. In some states, if the election is not successful, the bond issue cannot be put in front of the public again for a period of one year, unless the amount of the bonds has been decreased or increased by more than 5 percent.

THE BOND CONSULTANT

Negotiating a successful bond election is a complex process, one for which most school administrators are not trained and for which they do not have the time. A bond consultant not only guides the groups that will take the bond campaign story to the public, but he or she prepares the working financial documents and instructional documents necessary for the school campaign. Using a bond consultant ensures that the public will receive accurate information, that the proposed project will be realistic in terms of the district's actual needs and ability to levy additional taxes and allows administrators and board members to maintain a low profile in the election campaign—all factors that contribute to a successful bond referendum campaign. By utilizing a bond consultant, the board of education can meet all the legal requirements without the additional lawyer's fees.

The school board should adopt policies on capitalization programs (see Figure 3.1 for an example) and bond campaigns (see Figure 3.2 for an example). Such policies should reflect local and state guidelines for the bonding process.

To finance the facilities program, the Board, as established by law, may at its discretion authorize an annual tax levy not to exceed five mills on the taxable valuation of the district for the capital outlay fund. The Board may also issue and sell capital outlay certificates. Money received from the sale of these certificates will also be placed in the capital outlay fund.

The capital outlay fund is a fund provided by law for the purchase of land; improvement of grounds; construction of, additions to and remodeling of facilities; or for the purchase of equipment. It may also be used for installment payments for the purchase of real property, plant or equipment, where the installment contract does not exceed ten (10) years, and for the payment of the principal and interest of capital outlay certificates. When used for the purchase of capital outlay certificates and the payment of installment contracts, the total accumulated unpaid principal balances cannot exceed three (3) percent of the taxable valuation.

Construction of new facilities, or of additions to facilities which will require advertising for bids, must have a public hearing at least ten (10) days prior to the advertisement of any contract specifications. Following this public hearing and approval of the Board, the district may use the capital outlay fund for payment of the new construction or addition; however, the district may not change the originally advertised use of the fund without holding another public hearing.

In accordance with law, the Board will develop and maintain a five-year plan on the annual projected revenues and expenditures for the capital outlay fund. The projected expenditures will itemize the projected costs for new or additional facilities.

Established by Law.

(Adoption date, May 10, 1992)

Legal Refs.: SDCL 13-6-6 through 13-16-93

Source: Brandon Valley School District 49-2, Brandon, SD. Used by permission.

Figure 3.1. *Sample board policy—facilities capitalization program.*

In accordance with law, the Board by resolution may determine that the district should issue negotiable bonds. These bonds may only be used for the purposes of:

1. Refunding any bonded indebtedness which is or is about to become due and payable or whenever such indebtedness can be refunded at a lower rate of interest to fund any judgment or outstanding warrants; and

2. Raising money for any purpose for which the Board is authorized to spend school district funds.

The proposition to issue bonds, except bonds to fund registered warrants or to refund bonded indebtedness, will first be submitted to the electors of the district at a general or special election.

Established by Law.

(Adoption date, May 10, 1981)

Legal Refs.: Constitution of the state of South Dakota, Art. XIII, sec. 4.

Effective July 1, 1984 – Once the Board determines the necessity for a bond issue, the Board must obtain the services of a bonding company.

Source: Source: Brandon Valley School District 49-2, Brandon, SD. Used by permission.

Figure 3.2. *Sample board policy—bond campaigns.*

BONDING CONSULTING SERVICES

Bonding underwriters make their money from the sale of bonds to finance all sorts of building projects, mostly for government agencies. They, therefore, have a vested interest in assisting school districts to run successful bond referendum campaigns. A fiscal agent/bond underwriter should provide several types of services to bring about positive results. The agent/bond underwriter should:

1. Analyze the scope of the project, conduct a cash flow analysis, and explore financing alternatives in order to develop a well-documented financing plan that will be conducive to voter acceptance.
2. Coordinate the bond election information campaign by developing a pertinent timetable of events by organizing citizen committees through the assignment of clearly stated objectives, by preparing fact sheets and election brochures for public distribution, and by representing the school district at committee and public meetings.
3. Oversee the preparation of all necessary legal proceedings with a nationally recognized bond attorney.

4. Market the district's debt obligations.
5. Provide the printing and registration of the bonds, handling the closing and recommending effective investment alternatives for bond proceeds.
6. Monitor the district's outstanding debt obligations for potential refunding opportunities that will reduce interest costs and tax requirements.[4]

The school superintendent is responsible for obtaining the relevant materials from each investment firm under consideration. The bonding companies should be able to describe how they provide the services listed above, references, the number of and kind of public meetings the firm is willing to conduct, and answers to questions concerning the sale of bonds, timing of the elections, and their methods of organization.

Opposition groups often make an issue of the amount of money the board of education spends on passing the bond issue; therefore, it is important that the district utilize services that are the least expensive for the quality of service the school district needs. Bonding consultant firms generally represent one of three systems of operation. The first charges a fee for consulting, often including out-of-pocket expenses for the consultant.

Those firms that employ the second type of system do not charge a consulting fee, but they require the school district to cover out-of-pocket expenses for the actual consultant. Generally, the bond consultant will waive the fee if the bonds are obtained through their particular financial firm after a successful campaign.

Other bonding firms offer still a third arrangement. The firms indicate clearly to the board of education that there will be no fee for the consulting process if the bonds are purchased through their company. They do not charge for out-of-pocket expenses because they obtain their fees through the sale of bonds on behalf of the school district.

This last option is, of course, the preferred option because it keeps costs to a minimum; however, it should not be the only determining factor in selecting a bonding firm. The school administrator should discuss selection of a bonding company with others who have used the firm, preferably state officials, and local business people.

Once the school board has hired a bonding consultant, his or her first responsibility is to develop a business plan for the school district that includes a reinvestment schedule and cash flow analysis for the school district (see Table 3.1 for an example). The analysis should entail an

Table 3.1. Reinvestment schedule and cash flow analysis.

Reinvestment Schedule and Cash Flow Analysis—Any School District

G. O. Bonds = 4,725,000
Issuance Expenses = 94,500
Net Bond Proceeds = 4,630,500

Capital Outlay Certificates = 1,300,000
Issuance Expenses = 32,500
Net Proceeds = 1,267,500

End of Month	Construction Account Balance	Construction Draw	Bond Payments	Tax* Collections	Net Construction Account Balance	Interest** Income 8.25%	Investment Fund Balance
Feb. 91	5,898,000	0	0	0	5,898,000	40,549	5,938,549
Mar. 91	5,938,549	0	0	0	5,938,549	40,828	5,979,376
Apr. 91	5,979,376	0	0	0	5,979,376	41,108	6,020,484
May 91	6,020,484	115,600	0	0	5,904,884	40,596	5,945,481
Jun. 91	5,945,481	173,400	0	0	5,772,081	39,683	5,811,764
Jul. 91	5,811,764	289,000	0	0	5,522,764	37,969	5,560,733
Aug. 91	5,560,733	289,000	0	0	5,271,733	36,243	5,307,976
Sep. 91	5,307,976	578,000	0	0	4,729,976	32,519	4,762,494
Oct. 91	4,762,494	578,000	0	0	4,184,494	28,768	4,213,263
Nov. 91	4,213,263	578,000	0	0	3,635,263	24,992	3,660,255
Dec. 91	3,660,255	917,000	0	0	2,743,255	18,860	2,762,115
Jan. 92	2,762,115	917,000	0	0	1,845,115	12,685	1,857,800
Feb. 92	1,857,800	578,000	435,860	0	843,940	5,802	849,742
Mar. 92	849,742	289,000	0	0	560,742	3,855	564,597
Apr. 92	564,597	289,000	0	0	275,597	1,895	277,492
May 92	277,492	173,400	0	289,350	393,442	2,705	396,147
Jun. 92	396,147	115,600	0	0	280,547	0	280,547
Jul. 92	280,547	0	0	0	280,547	0	280,547

(continued)

Table 3.1. (continued).

Reinvestment Schedule and Cash Flow Analysis—Any School District

G. O. Bonds = 4,725,000	Capital Outlay Certificates = 1,300,000
Issuance Expenses = 94,500	Issuance Expenses = 32,500
Net Bond Proceeds = 4,630,500	Net Proceeds = 1,267,500

End of Month	Construction Account Balance	Construction Draw	Bond Payments	Tax* Collections	Net Construction Account Balance	Interest** Income 8.25%	Investment Fund Balance
Aug. 92	280,547	0	217,930	0	62,617	0	62,617
Sep. 92	62,617	0	0	0	62,617	0	62,617
Oct. 92	62,627	0	0	0	62,617	0	62,617
Nov. 92	62,617	0	0	289,350	351,967	0	351,967
Dec. 92	351,967	0	0	0	351,967	0	351,967
Jan. 93	351,967	0	0	0	351,967	0	351,967
Feb. 93	351,967	0	347,930	0	4,037	0	4,037
Mar. 93	4,037	0	0	0	4,037	0	4,037
Apr. 93	4,037	0	0	0	4,037	0	4,037
May 93	4,037	0	0	289,350	293,387	0	293,387
Jun. 93	293,387	0	0	0	293,387	0	293,387
Jul. 93	293,387	0	0	0	293,387	0	293,387
Aug. 93	293,387	0	212,740	0	80,647	0	80,647
		5,880,000	1,214,460	868,050		409,057	

*Payments from tax collections are based on the average amount of payment required each fiscal year for the bond issue with a 98% tax collection rate.

**Interest income computed on the month end balance of the construction account and the bond fund. All interest income is deposited into the construction fund. (The net construction account balance times the annual interest rate divided by 12 months equals the monthly interest amount.)

The reinvestment schedule and cash flow analysis example was provided by Darwin Reider, First Vice President, Kirkpatrick, Pettis, Smith, Polian, Inc., Omaha, Nebraska.

Months of	
February/March	School board, school administration, architect, and fiscal agent determine scope of building improvement program.
Day of	
March 13	School board passes resolution to call an election.
March 15	Mail letter of invitation to potential committee volunteers for March 21 organizational meeting.
March 16	First publication of Notice of Voter Registration.
March 21	Organizational meeting for sub-committees.
March 23	First publication of Notice of Election. Second publication of Notice of Voter Registration (at least 10 days before deadline).
Week of	
March 27	Sub-committees meet on their own to discuss strategies to accomplish stated objectives.
March 30	Second and final publication of Notice of Election (20 days before election). Publication of sample ballot.
Week of	
April 3	Sub-committees begin dissemination of informative materials (fact sheets and general information brochure to designated neighborhoods).
April 10	Deadline for voter registration (15 days prior to election).
April 15 and 20	Public information meetings (if necessary).
April 25	School bond election. Board meets to canvas election results.
May 1	Board passes resolution to issue bonds (if market is favorable).
June 1	Closing of issue and reinvestment of proceeds.

Figure 3.3. *Sample—election timetable, any school district.*

in-depth analysis of the assessment values of the district, a determination of the growth potential of the community, and other resources (capital outlay certificates, capital improvement fund balances, and possible gifts for the facility) that are available. From this study, the bond consultant can make a fairly precise estimate of the financial capabilities of the district and what type of tax increases are realistic. This analysis is critical in that the amount of new taxes being levied is a significant factor in the success or failure of bond elections (see Chapter 1).

The second task for the bond consultant is the development of a possible election timetable as shown in Figure 3.3. The timetable should clearly indicate time lines that must be followed to take the election campaign from start to finish. The consultant must take into consideration the immediacy of the need, the tiring factor for volunteers, and the relatively short attention span of the general public.

By the time the bond consultant has finished his or her preliminary work, the school administrator and board of education should have a fairly clear picture of the situation they face. Most important, they should understand how much money they can realistically spend on a renovation or construction project.

SELECTING THE ARCHITECT

At the same time the school superintendent is determining which bond consultant to recommend to the school board, he or she should be conducting a search for an architect for the project. Once again, the board should adopt a policy that delineates the selection process criteria. The policy may also establish criteria for determining when they do not need to hire an architect. The criteria[5] for selecting an architect are as follows:

1. Experience in school design
2. Evidence of relevant experience in special situations, such as facilities for the handicapped
3. Creative design ability
4. Technical knowledge to control the design so that the best results are obtained for the least amount of money
5. Executive and business ability to oversee the proper performance of contracts
6. Proven ability in all the major phases of planning and construction:

predesign planning, schematic design, design development, bidding, and construction
7. Ability and temperament to work cooperatively with others
8. Willingness to consult with staff on educational specifications
9. Extent and experience of architectural staff in relation to the scope of the planned project
10. Reasonableness of architect's fee[6]

The first step for the school superintendent is designing and distributing a search document. The superintendent should send such a document to all the architects in the region and architectural firms outside the region who have worked previously in the area. The document should include a brief description of the type of facility the district wants to build, the criteria for selection, and the time lines for both the selection process and building project. Also, the letter should contain information as to how the architect can apply for consideration.

To ensure that no architect is missed accidentally, the superintendent should place a notice in at least the local newspaper indicating that the school district is conducting a search for an architect. This notice should include the information listed above.

Depending on the number of applicants, the school superintendent should narrow the number of applicants to be interviewed by the board to three or four. The superintendent should ask each of the candidates to prepare a presentation that portrays his or her experience, abilities, and ideas on educational facilities. They should also be prepared to discuss their qualifications under the 10 criteria listed previously, including concessions they might be willing to make in their fee schedule. For example, some firms will be willing to do all of the preliminary drawings, preparation meetings, and training programs necessary for the bond campaign at no cost to the school district until the voters have approved the building project. In some states, laws and regulations determine fee schedules, and out-of-state firms must be willing to comply with those restrictions.

The superintendent should schedule interviews for from 45 minutes to an hour. It is extremely important that all school board members be involved in the selection process and that they make every attempt to reach a consensus. This brings the benefit of another unanimous vote by the board, an important element in achieving success in the bond referendum election (see Chapter 1).

After the board has hired an architect, the superintendent should reach agreement with the architect as to the type of contract, specific fee schedules, and cost controls on change orders. The board of education should adopt a policy that states that contracts with the architect must meet the current standards of the American Institute of Architects. Following this standard can eliminate many problems of personal interest. School superintendents also should use the estimated costs of the project as a starting point for developing an appropriate fee schedule. The architect may also be willing to "lock" his or her fee, so change orders occurring during the building project would not fall under the architect's fee schedule. This helps eliminate concerns among the public that the architect intentionally forgot some items or that he or she built in some errors to increase his or her fees.

After the hiring process is complete, the superintendent should contact all applicants thanking them for their interest, informing them of the firm that was hired, and wishing them the best in the future. Some firms may contact the school superintendent as to the reasons the board did not select their firm, so the superintendent should be prepared to identify a couple of points such as differences in presentation, ideas, or finances.

EDUCATIONAL SPECIFICATIONS

The school district is responsible for providing the architect with a set of educational specifications upon which he or she is to base the building's design. The specifications should be developed from input provided by educational consultants, teachers, administrators, and parents. Information from the school plant data sheets (see Chapter 2) can also be used.

Educational specifications can be divided into three basic categories. The first is devoted to a detailed description of the educational activities that will take place in the facility. This is not an expression of educational philosophy; it is a description of what people will be doing in various sections of the building (e.g., the lobby, the music classroom, the physics lab, and so on). It should also include descriptions of the dimensions and purpose of certain types of equipment used in some classroom spaces and service areas, and what activities take place during certain parts of the year (the architect needs to include space for three activities going on in December and he or she has only provided space for two, even though during the rest of the year only two activities are in progress).[7]

The second category is physical specifics of the facility. Information needed includes how many students will attend the school, potential increases in enrollments, how many students are typically in a certain class at the same time, how many spaces are needed for each time slot in a day, and so on. Information in this category should also include space relationships. For example, it seems logical that the chemical storage room should be located next to the chemistry classroom, but most non-educators would not consider the importance of having the door to the guidance center separate from the door to the main office.[8]

The third category includes a description of special physical features the school program demands. These considerations often are concerned with special educational program demands. Special features include requirements for room shape (e.g., home economics), ceiling height (some vocational classrooms), intensity of lighting (art), humidity and temperature control (e.g., science laboratories), color, ability of floor to support more than average weight (e.g., libraries, stages), built-in equipment (e.g., video systems), and so on.[9]

All three categories should include provisions for special needs students. Architects usually have references that list standards for the industry, building codes, and square footage requirements. However, school officials should never "leave it to the architect." Estimates of the total amount of classroom space needed, the space required for the gymnasium, and other considerations should be spelled out specifically for the architect.

It should be noted that the type of specifications addressed by the architect will vary depending on whether the project is a major renovation or a new building project. It also should be noted that state departments of education often have guidelines as to minimum acreage, minimum design standards, and minimum materials standards for public school structures. The school superintendent should be certain that the architect is aware of such requirements.

In addition, the administration should not ignore requirements for the basic esthetic appeal of the structure. If this element is forgotten, the district might end up with a brick box. The school superintendent should provide the architect with some guidelines as to the decorative aspects of the building. What is the basic attitude of the school district in terms of esthetics (that is, Would they find an avante-garde design objectionable?)? What materials would be acceptable to the community (i.e., Are they going to expect a traditional brick building or will they accept a mixed material exterior?)? What interior designs and colors would students and teachers find most comfortable and inviting? Are there any historical considerations? What types of residences are located in the area in which the school will be located? The importance of an appealing building should not be underplayed—it might be the reason a bond issue fails.[10]

PRELIMINARY DESIGNS AND OTHER CONSIDERATIONS

The next task of the architect is the development of the preliminary design. The school superintendent and school board members should examine the design carefully before they present it to the public. The critical issue at this time is insuring that the criteria for the design (i.e., needs of the school district and educational specifications) have been met. The architect also should be able to present the board with preliminary estimates regarding construction costs and job completion time lines.

The architect may wish to indicate the type of information he or she should provide the public concerning the design. Such information should include what the architect determined were priorities for the structure and the design features that accommodate those needs, room arrangement and design elements that provide the most efficient use of space and most comfortable traffic patterns, and esthetic features and why they are important to the overall effectiveness of the building, design features that provide space for integration of educational programs and use of technology, and design features that provide ease of accessibility for all students.

The architect holds a great share of responsibility for explaining the building design and its advantages over other designs. He or she is the individual with the expertise to answer questions about how the new structure will be constructed and what unique features the building has. His or her involvement in the bond referendum campaign is vital to the success of the campaign in that it provides the public with access to accurate information and, once again, allows the superintendent and board members to maintain a low profile. The board should develop an understanding with the architect early on about the extent they wish him or her to participate in public presentations, training sessions for volunteers, and radio or television appearances. While some architectural businesses provide a certain amount of assistance in this area, many charge for expenses. The school board and architect should reach a clear understanding about the financial obligations of each before the architect is asked to participate.

The architect supervises certain aspects of the actual construction of the facility and is responsible for obtaining the appropriate bids from suppliers and subcontractors. The school board may legitimately expect that the architect will keep the project on budget as much as possible, without making serious sacrifices in terms of the quality of materials or specific design features. The architect should submit any requests for changes to the school superintendent and the school board prior to enacting such changes.

REVIEW ACTIVITIES

1. Make a list of all repair, renovation, or building projects in your district that may require capital outlay expenditures during the next five years.
2. Examine the spending practices of your school district and determine which of the above can be financed through the general fund or capital outlay fund.
3. From the research findings above, decide what projects probably would require a bond issue.
4. Make a list of bonding companies in your region that have experience handling bonds for educational institutions. Summarize and compare the policies and the services each provides.
5. Make a list of architectural firms in your region that have experience in designing educational facilities, and summarize and compare their policies and the services each provides.

ENDNOTES

1 Meierhenry, T. 1997. "South Dakota School District Bonds: A Guide to Public Indebtedness." Prepared by Danforth, Meierhenry & Meierhenry, LLP, Sioux Falls, SD.

2 A list of municipal bond attorneys is published in *The Bond Buyer's Municipal*

Marketplace, formerly the *Bond Buyer's Directory of Municipal Bond Dealers*, commonly referred to as the "Red Book."

3 National Association of Bond Lawyers. 1988. "Selection and Evaluation of Bond Counsel."

4 Reider, D. 1993. "Why Does a School District Need a Fiscal Agent/Bond Underwriter?" Omaha, NE: Kirkpatrick, Pettis, Smith, Polian, Inc.

5 ———.

6 Castoldi, B. 1994. *Educational Facilities: Planning, Modernization and Management*. Fourth edition. Boston: Allyn & Bacon.

7 ———.

8 Earthman, G. I. 1992. *Planning Educational Facilities for the Next Century*. Reston, VA: Association of School Business Officials International, pp. 140–149.

9 Castoldi, B. 1994. *Educational Facilities: Planning, Modernization and Management*. Fourth edition. Boston: Allyn & Bacon.

10 Holt, C. R. 1993. "Factors Affecting the Outcomes of School Bond Elections in South Dakota," Ed.D. diss., University of South Dakota, pp. 9–50.

CHAPTER 4

The Project Development Phase: Bringing It All Together

TO many it might appear that the information that has been gathered is scattered among several people and groups. The bond consultant has been collecting and analyzing the financial information; the architect has gathered information about the needs of the school; the community group has gathered information about community attitudes and support; and the school superintendent has been gaining input from school staff and administrators and individuals within the community about what the community needs and wants.

At this point the board of education needs to appoint committees whose function will be to bring together all relevant information, analyze the data, and make final decisions as to the site, building design, financial obligation, and so on. These committees, which may be appointed and funded by the board of education, prepare the final recommendations to the board. Because these committees are involved in development rather than promotional activities, membership on the committees is not limited to persons living in the district. Indeed, it probably is advisable to have some school administrators, teachers, and support staff members serve on the committees.

Generally, the committees that will be needed are: (1) a steering committee, (2) a site committee, (3) a structure committee, (4) an educational obsolescence task force, (5) a community uses task force, and (6) a finance options task force. The responsibilities of each committee will be discussed later in the chapter.

Finding people to serve on committees is never an easy task, and busy people tend to go the route of least resistance when asking people to serve: in other words, they go to the people they know will serve. Such actions, however, may cause segments of the community to perceive the entire process as one of a group of community "elites" trying to push something down everyone else's throats. They might even believe that they are not being provided the whole story and that decisions are being made in secret. Any of these perceptions can spell disaster for a bond referendum campaign.[1] The community leader and school superintendent, who are in charge of forming the committees, must make every effort to include a cross section of the community on each committee. As much as possible, selections for the committees should include representatives from the various socioeconomic, ethnic, religious, and political groups represented in the community. The membership also should be made up of people of various ages and with a diverse set of interests. While it might not be appropriate to select individuals who are strongly opposed to the project, the community leader should attempt to put at least a few persons on committees who are undecided about the bond issue.

FORMING THE COMMITTEES

By this time the community leader and school superintendent should have a list of people from the district who either support or are leaning toward supporting the proposed building project. From that list, the community leader and superintendent may select a large group of individuals who will be asked to serve on committees. These individuals should be contacted about their willingness to attend a meeting where participating on committees is to be discussed. The community leader should contact each person on the list to ask them to attend this special meeting, informing them of the time and place.

The purposes of this initial meeting are:

1. To disseminate information as to the need for the building project, the scope and design of the project, and the estimated costs of the project
2. To inform the participants about the purpose and duties of each committee, the time lines for work, and what would be expected of each committee member in terms of duties and time investments
3. To provide an opportunity for individuals to give input as to their ideas about the project

While the community lead person should conduct the meeting, the school superintendent, bonding consultant, and architect should be available to provide specific kinds of information as needed. At this point, all individuals involved in leadership positions should maintain an objective stance, sticking to a presentation of facts.

At the end of the meeting, the group should be asked to think about whether they would be willing to serve on a committee, which committee they would prefer, and what the level of their commitment could be. The community leader should point out that not everyone on a committee has to be a leader (translation: not everyone will have to speak in public), and that people who are willing to work behind the scenes are just as important as those who are "out in front." The community leader needs to establish a deadline for people to sign up for committees, usually from two to four days.

During the next few days, the community leader should contact all individuals who attended the meeting to encourage them to participate on a committee. After individuals have made a commitment, the community leader needs to set up meeting times for all committees.

THE STEERING COMMITTEE

The steering committee has as its mission identifying and coordinating necessary tasks leading up to the formulation of the formal proposal to the board. The committee has the following objectives:

1. To organize the information provided by the other committees into a final recommendation to the board of education
2. To prepare appropriate documents for the dissemination of information to the public
3. To recruit volunteers to distribute information, make phone calls, or perform other proposal development services

Resource persons for this committee include members of previous board committees, parents, the bond consultant, teachers, and administrators. One strategy for assignment to this committee is to include the school superintendent, the community leader, and the chairs of the site and building committees. A person skilled in writing and/or graphics might also be assigned to this committee.

This committee bears a great deal of responsibility and probably will have to meet at least every other week during the early stages of the development phase. The committee may have to meet more often as they begin to assemble the final report.

THE SITE COMMITTEE

The site committee's task is to make a recommendation to the board of education as to the most suitable location for the building project. The committee may have several options from which to choose, or they may have only a few. Those options might include building on land already owned by the school district, purchasing a more desirable portion of land, or trading school-owned land for more suitable land.[2] The site committee might also decide whether the facility should be in a new area or should be attached to an existing facility, although this decision would rest on the board's decision to renovate or build.

The committee examines the various options, considering the following criteria:

1. Traffic patterns in the area: These should include consideration of the safety and ease by which students may get to the school (walking, bicycling, and vehicles). Travel distances from residential areas should be investigated.
2. Availability of city services in the area: Are telephone, electricity, water, and sewer already in place? Are the roads adequate to handle the traffic? Do fire and police services have easy access to the site?
3. Bus routes: The committee should assess the distances traveled and the amount of time students would spend on the bus.
4. Site implications: The committee analyzes the following in terms of their implications for the building: site topography, natural windbreaks, groundwater elevation, surface drainage patterns, and soil bearing pressure. In most cases, the school board will hire specialists to take actual measurements and do the statistical analysis. The problem for the committee is analyzing the information in terms of the feasibility of building on a particular site.

   They should also address the implications of zoning ordinances in adjacent areas, accessibility for delivery and maintenance, the degree to which expansion could occur on the site, and problems of flood control. The committee will also have to determine any city, county, or state regulations that would affect a potential site or the ability to acquire the land if it is not already owned by the school.

5. Costs: The committee must analyze how the cost of obtaining the land fits into the overall anticipated budget for the entire project.

In a nutshell, the major task of this committee is to do a cost/benefit analysis of the various sites that could be available and make a recommendation to the school board. For example, a certain site may be in the proper location according to distance from population centers and traffic patterns; however, the topography would require added expense because large amounts of earth would have to be removed to create a level site. Another site might meet land requirements, but the district would have to bus students long distances to the school. The site committee must decide what the district's priorities are and which site meets those priorities most adequately.

Resource people for this committee include the architect, the bond consultant, the city planner, community safety officers, public utilities officers, construction engineers, and residents of surrounding areas.

This committee probably will operate most efficiently if it is broken into task groups that are responsible for analyzing certain types of data. For example, individuals with building experience might work most effectively with the architect on analyzing the site implications, while parent representatives would be most effective gathering information from parents about transportation problems.

The conclusions of the committee may be presented to the board of education in several ways. First, the committee can make a specific recommendation to the board, requesting they choose a specific site. The committee also might present a list of categories to the board and how they ranked each site alternative in that category. Finally, the committee might narrow the choices to the top two and then prepare an analysis of the choices.

The work of this committee is fairly technical and may require the members to do some research. The committee should probably anticipate spending from three to five weeks preparing their recommendations, depending on the number of sites they must review.

THE STRUCTURE COMMITTEE

This committee probably has the most daunting task of any of the committees, and those who serve on this committee should anticipate spending from four to six months accomplishing their task. The number and complexity of the tasks necessitate forming subcommittees to investigate some of the issues.

The major task of this committee is to make a recommendation to the board of education as to the most critical design specifications for the facility. The committee should base its recommendation on the following criteria:

1. The educational program of the school district: What types of facility (square footage, special wiring, layout, and so on) best meet the needs of the educational program the community wants? Such considerations should include the type and arrangement of rooms, features that have an impact on discipline and control, multi-use facilities, common areas (cafeteria, gymnasium, libraries, locker rooms), the number of classrooms and classroom capacity, and inclusions of advanced technologies.
2. The health and safety needs of the students and faculty: What are the federal, state, and local fire regulations, special needs regulations, and building materials codes?
3. The potential costs of maintenance, heating, or cooling
4. Potential community use: What are the possibilities for organizations in the community using the facility and what impact should this have on the final design?
5. Costs of various alternatives: What are the costs of the various alternatives and how do those costs fit into the overall financial conditions of the district?

To accomplish these tasks, the committee should break into several task groups.

Educational Obsolescence Task Force

This group reviews current and future instructional programming to identify limitations caused by current facilities and what features would need to be included in the new design to accommodate these programs. Their objectives should include, but may not be limited to:

1. Identifying the limitations caused by the current facility in the areas of reading, language arts, social studies, science, math, health, the arts, physical education, counseling, and special education
2. Identifying the limitations of the present facility in terms of new technologies in the classroom

3. Identifying the limitations of the present facility in terms of extra-curricular activities and student participation in those activities
4. Identifying the limitations of the present facility in terms of accessibility and health and safety concerns
5. Identifying the limitations to effective control and teaching (e.g., overcrowding) caused by the lack of space in the present facility
6. Identifying those elements of the facility that must be incorporated into the design to meet the minimum standards necessary for the enrollment in the school, educational programs, and extra-curricular activities

A great deal of this work will have been accomplished already through the facilities surveys and input from teachers, support staff, and members of the community. The committee should put this information together in a unified report to the larger committee.

Potential resource persons for this task group include the superintendent, architect, city engineer, teachers and support staff, and parents.

Community Uses Task Force

The purpose of this task group is to assess the facility needs of community groups, both public and private, that could be answered by design features in the new building. The committee also must evaluate ways in which the facility could be utilized by the general public and what elements of design would need to be modified to meet those uses. This item is an important one for several reasons; the most important being that the percentage of voters needed to pass a bond issue may be reduced to 50 percent in some states if the facility is built jointly by the school district and another agency (thus the reason so many school gymnasiums are in National Guard armories).

The tasks of this group include:

1. Contacting various organizations and groups that might want to use the school's facilities in the future and identify what their needs would be (e.g., size of rooms needed, special equipment needed, and so on)
2. Identifying any state or federal regulations that restrict the use of public property by private or religious groups
3. Identifying the current board of education policy and practice on utilization of school property by private and religious groups

4. Identifying possible cost-sharing possibilities and what organizations might be willing to do
5. Preparing a report to be presented to the larger committee

Potential resource persons for this group include the architect, city manager, representatives of senior citizen groups, representatives from the city council, county officials, National Guard officials, and religious leaders.

The task group should plan to be active for six to eight weeks.

Finance Options Task Force

The purpose of this task force is to develop a comprehensive, understandable explanation of the finance options for capital improvement; the effects on local sources (property taxes); and the short- and long-term effect of new construction versus maintaining the status quo and analyzing the financial advantages/disadvantages of building designs to determine which would be the most cost effective. The work of this committee is crucial to passage of the bond issue because its studies represent the justification for inclusion of certain design elements in terms of long-term costs of the facility. For example, how long will it be before thermal windows pay for themselves by reducing fuel expenditures?

The major tasks facing this committee include:

1. Reviewing sources of revenue currently available for total district operations
2. Reviewing sources of revenue available for capital projects
3. Identifying the most appropriate finance option for new construction
4. Comparing the costs of maintaining and operating the present facility to the costs of maintaining and operating new structures
5. Providing comparative analysis of cost effectiveness of alternatives in a new building such as type of windows, type of heating and cooling systems, building materials, square footage for certain areas, and cafeteria systems
6. Reviewing effects of tax hike in relation to trends in tax assessments over the next five to ten years
7. Developing a written report for the school board, media and general public

Potential resource persons for this committee include the school business manager, school administrators, representatives from financial institutions, building supply representatives, county assessor, and the bond consultant. This committee will gather a great deal of data and compile the information so that the advantages and disadvantages of alternatives can be identified easily. Committee members should anticipate working from five to eight weeks on the project.

This committee also must keep in mind that voters will be deciding whether or not they like the facility or want to pay for part or all of it. In his article, "How to Blow a Bond Issue—Or Not, If You'd Prefer,"[3] Carter makes some cogent suggestions on some key elements to consider before making the final recommendation.

First, Carter suggests that trying to "pork barrel" the building is not an effective strategy. This set of actions calls for the building to include features that everybody knows are needed (e.g., handicap accessibility) and then a series of extras that are preferred by one special interest group or another. For example, some people believe the big attraction to the schools for non-parents is the athletic program. Therefore, they believe the plans should include an emphasis on those areas. These activities are also more exciting than video systems or an English classroom. Carter postulates that many school boards believe that, even though some people don't care for some of the items included, they will care enough about the ones they do like to vote for the bond issue. He does not believe such is true.

The same article enjoins school boards to "keep your eyes on the prize: if you don't need it, don't ask for it." Carter suggests that one of the reasons bond issues fail is that the building is designed with facilities that the ordinary taxpayer cannot see make a difference to education. For example, more classrooms are a good thing; bigger parking lots probably are not. Expanding the library is a good thing; expanding administrative office space is probably not. Stick to those elements of the school design that are for kids.

Carter's final suggestion is to really listen to what the voters are saying. If the plan outlined in this book is followed, the superintendent and other officials will have a great deal of information about what district patrons want. Their ideas should be worked into the plan. Voters may revolt if they believe the proponents of the bond issue "just asked my opinion to be nice."[4]

THE SCHOOL BOARD RESOLUTION

During the above process, each of the committees write reports to the steering committee outlining their work, giving their analyses, and giving their final recommendations (see Chapter 8 for information on preparing these reports). The steering committee then works to make a unified presentation to the board of education outlining the specifics of the project they would recommend. The end result should be their recommendation for a specific type of school, located at a specific site, at a specific cost.

Those who go on to work for passage of the bond issue referendum should use the materials created by these committees. The research, statistics, and opinions gathered for this phase prove invaluable in explaining the justifications for expenditures. They can also help the administration and others understand what alternatives might be available to them should the initial bond issue fail.

After the steering committee has made its final recommendation to the board of education, the board then votes either "yes" or "no" to a resolution calling for a bond referendum election to take place at a certain time in the future. Opinions vary as to how long after the bond resolution is passed that the election should be held, but most experts caution about letting it slide too long. The bond resolution should contain naming the specific project, the location of the project, a maximum cost expectation, and the date and time of the election. In many states there are specific regulations governing how the resolution must be worded, what statements must be made, and so on. The school attorney should prepare the resolution carefully following state guidelines.

The process is now reaching its final phase: the election campaign.

REVIEW ACTIVITIES

1. Make a list of individuals within your community who might serve as community leaders, resource persons, volunteers, or committee members.
2. Study state laws that might affect how the resolution should be stated.
3. Develop a list of tasks for each committee: Steering Committee, Site Committee, Structure Committee, Educational Obsolescence Task

Force, Community Uses Task Force, and Finance Options Task Force.

ENDNOTES

1 Carter, M. A. 1995. "How to Blow a Bond Issue—Or Not, If You'd Prefer." *Clearing House* 68(5): 289–292.

2 Castoldi, B. 1994. *Educational Facilities: Planning, Modernization, and Management.* Fourth edition. Boston: Allyn & Bacon.

3 Carter, M. A. 1995. "How to Blow a Bond Issue—Or Not, If You'd Prefer." *Clearing House* 68(5): 289–292.

4 ———. pp. 289–292.

CHAPTER 5

Beginning the Campaign: Developing a Marketing Philosophy

AFTER submitting their reports to the board, the steering committee is largely finished with its work, but after the board makes the decision to proceed to the bond election, the public relations committee is just beginning theirs. This committee is responsible for organizing the effort to get the correct information to the public and for organizing efforts to pass the bond referendum resolution.

The campaign is, in effect, a marketing effort, and the campaign planners should follow the basic principles of any marketing strategy:

- Know the product, both what it is and what it isn't.
- Know the target population, including their needs, wishes, and core beliefs.
- Know the best marketing techniques to get the message across to that group.

KNOW THE PRODUCT

So far, the committees and others have generated most of the information about the new facility needed by the public relations committee to present the proposal to district patrons. One must be careful here to insure that the product being sold is a quality education—not a specific building. Almost every individual within a school district wants the district's children to get the best education possible. If they are not parents, they are probably grandparents, and for the most part they care about young people. But they want to know how their tax dollars will contribute to education: they want to know they will not be throwing their money away.

Proponents of the school bond referendum must be able to justify almost any expenditure in terms of how it will help provide a better future for the district's children. Why does the school need computer labs? Why is assuring access for disabled persons so important (besides the fact that the federal government mandates it)? How does building a new gymnasium for extracurricular activities help students learn better? Patrons of the district will ask these types of questions, and the public relations committee should have the answers.

Voters also will want to know how educating young people contributes to their quality of life. They may be able to identify that extracurricular activities can provide entertainment, but they may not see how adding space to a library for computers helps them. They must be given information that points to how good schools attract businesses to a community, thereby increasing the tax base and making property taxes lower for everyone. Someone must show them that providing a pleasant environment in which students learn contributes to those students having a more positive attitude toward the community, and how that, in turn, might contribute to a lower juvenile crime rate and lowered drug use.

The above analysis points to several important strategies for public relations committees:

1. The campaign must be about education, not about square feet, new lockers, or art rooms.[1]
2. The public must be assured that the management of education in their community is solid and rational. Almost everyone knows stories about businesses that had beautiful facilities, but because of poor management or poor goals, went bankrupt. The public must have confidence in the management of education in their community and then be shown how the new facility fits into that management plan.[2]
3. The campaign must focus on how education benefits the community and all individuals within that community.[3]

KNOW THE TARGET MARKET

The targeted group in this marketing campaign is those persons who will vote in the election.[4] Some people believe that if they can convince the parents within the district, the election is won. Nothing is farther from the truth. In most districts, if every parent in the school district voted "yes" and no one else did, parents would not constitute a large enough number to even come close to passing the issue.

Some individuals also believe that the people who show up at public meetings represent the people who will vote in the election: again, an inaccurate perception. In fact, according to observations by persons researching school bond elections, on average, fewer than half of the people who attend public meetings live in the district or are registered to vote there. Many persons who have a personal stake in the referendum (real estate agents, developers, architects, contractors) attend the meetings to take advantage of the type of business information they can get. Naturally, they are mostly in favor of the bond referendum's passage because it would make a positive contribution to their welfare.[5] To determine who the market consists of one must analyze which groups participate in elections.

Statistics from voter participation surveys in presidential and congressional elections in 1988 and 1992 provide valuable information. That data clearly indicate that the older segments of the population are the ones who vote most consistently. In the two elections, almost 70 percent of those 45 years old and older who were registered actually voted in the election, compared to only 36 percent of voters between the ages of 18 and 45.[6] There is little reason to believe that the demographics for participation in bond elections are any different (although the turnout in general for bond issues is much lower). Unfortunately, those in the community with the most to gain from school building projects, namely parents, fall into the age groups which typically vote the least often.

The above would indicate two important strategies for the public relations committee:

1. To find ways to appeal to senior citizen groups
2. To conduct voter registration and "get out to vote" campaigns targeted at persons between the ages of 18 and 45

Voters within the community can be classified into four groups: those who are absolutely opposed to building anything, those who are undecided but leaning toward voting "no," those who are undecided but leaning toward voting "yes," and those who support the bond issue. Individuals within these four categories do not necessarily fall into any specific age, socio-economic, or political group. For example, many senior citizens are very supportive of programs to educate young people, while some young people (even those with children) object to any effort to increase taxes. Usually, property owners take more interest in the bond election than do renters, but that doesn't mean that property owners will be more skeptical than renters when it comes to voting in the election. In his book, *Educational Facilities: Planning, Modernizing, and Managing,*[7] Castoldi identifies attitudes held by individuals in the above mentioned categories.

The group that is supportive of the bond issue typically sees the relevance of education to their lives and to the welfare of the community as a whole. They see a good education as necessary for an improved quality of life to themselves.

Those who are undecided but leaning toward voting "yes" generally see the benefits of education to young people, but are uncertain as to the benefits of this particular project or believe an alternative solution might be better. They may object to the site that has been chosen. Generally, this group understands the need, but is not sure how best to fulfill that need. For example, some people might understand that the school district's elementary school is not adequate for its present enrollment and that the school district has to do something, but they have questions about whether a new structure should be built or an addition should be built onto the present facility.[8]

Those who are undecided but leaning toward voting "no" may also have more questions about the proposed project than about the need for a good education. These individuals may have serious objections to parts of the project they consider "frills" or luxuries. These individuals are concerned about the cost, but they all do not object to raising property taxes to pay for things that are truly needed. For example, many lean toward opposing a school bond issue when schools start talking about swimming pools, weight rooms for athletes, or theaters for drama groups.

Those who are absolutely opposed to the bond issue generally are those who would object to any project that would raise their property taxes. Although some people do not see the importance of education, most people who make up the opposition have reasons that are not as malevolent as many who support the bond issue might like to think. Older citizens on fixed incomes and younger people having difficulty living from one paycheck to another have real concerns about being able to hold on to their homes. Some have an anti-government attitude that is not directed at the schools. Some of the anecdotal information they receive may give them good reason to question some programs. For example, in one small, rural district in South Dakota compliance with the special education program a parent demanded for her child, who moved into the district, would have increased property taxes for each property owner by almost $1,000 a year. Put enough of these stories together and the whole educational system becomes suspect.

An analysis of the attitudes of each group points to several strategies that the community relations committee should consider:

1. The campaign should be a positive one. It must focus on priorities that the community set during the input phases of the process and how the new facility meets those needs. Ridiculing or putting down those who oppose the bond issue only solidifies that group and tends to draw "undecided" voters toward the opposition. Respect the opinions of everyone.
2. The campaign should emphasize the cost/benefit ratio of the various parts of the facility design. For example, in a fictitious bond issue many citizens have voiced objections to building an expensive theater in the new high school. The committee should not downplay the cost of this feature; instead, they should show how the facility will "pay for itself." The school would not have the expense of renting other facilities for its plays; community groups could now bring in speakers, concerts, and other cultural activities that require a modern and large facility; and the school district could now host play contests, some of which would bring two to three hundred people into town.
3. The costs of providing facilities for mandated programs should emphasize the school district's commitment to such programs. It only lends to negativity to say, "Well, we wouldn't have included that, but federal law says we have to. We don't have a choice." The opposition will not be placated by blaming others.
4. The group should emphasize that the administration and board took a conservative approach to the building design. This can be accomplished by giving indications as to what other school districts have done that would have cost more or taken more land. The committee must continually come back to the theme that the school administration and school board are looking out for the best interests of the community.

KNOW THE APPROPRIATE MARKETING TECHNIQUES

Several underlying philosophies must guide the committee in determining which marketing techniques they will employ in the bond referendum campaign:

1. Voters want the facts of the issue, unclouded by broad emotional appeals. Therefore, the committee must adopt an attitude of openness and respect. Trying to manipulate information to get voter support, rather than dealing with what is actually the case, is a sure course to disaster. Some proponents confuse fancy brochures and video presentations with keeping the voters informed about how much the bond issue will cost them and details of the planned facility.

   Some campaigns employ scare tactics rather than honest answers. The public relations committee in one school that eventually lost the election sent brochures home with the children, detailing to parents how their children would receive only a second-rate education if they did not vote for the bond issue. The brochure was filled with language detailing the poor fate that would befall the students.[9] Law enforcement officials engaged in the war on drugs have discovered that scare tactics usually don't work, and the same applies to bond elections. Indeed, many people resent such tactics as being condescending and childish.[10]

   The public relations committee should be disseminating concrete information, not philosophical educational jargon. Those who present materials in public should avoid invoking "educational experts" opinions as their means of defense. Rather than point to a national problem, the presenters should point to local problems and the ways to solve these (e.g., the national dropout rate doesn't mean as much to laypeople as the local dropout rate). While emotional appeals can and probably should be a part of the campaign, they should play a secondary role to the dissemination of information with which people can identify.[11]
2. The committee also must take into consideration how people develop their attitudes toward schools.[12] Most people make judgments based on what they personally have experienced or what they personally have observed. For example, older people grew up during the Depression, when educational opportunities were limited. They also perceive themselves as happy, successful individuals. They find it diffi-

cult to understand why education has to be so complicated today. They don't trust all the new innovations (e.g., "computers will ruin our world"). Likewise, most older people grew up during a time when discipline was the hallmark of the schools. They see young people hanging around the school and get a negative perception of how the schools are run. They question whether the school administrators know what they are doing. They find it difficult to understand why the law guarantees students such an array of freedoms. The public relations task becomes one of selling the management of the schools.

3. The information delivery mediums used should be those avenues through which the public believes they receive the most and best information. In a study prepared for the American Association of School Administrators in 1994, Mellman-Lazarus-Lake determined that most people believe they get a majority of their information about the schools from newspapers (30%) and friends and children (21%). The interesting result is that those surveyed placed one-on-one sources much higher than all media in terms of the quality of information they provide. Participants in the study gave the most "very good" ratings to school principals (37%) and the lower number of "very good" ratings to school superintendents (29%); however, when the percentages for the "very good" and "good" categories are combined, the highest ratings go to parents (80%) and the head of the local parent organization (80%). Local newspapers were rated in the "good" and "very good" categories by only 63 percent of the respondents.[13]

The above data indicate that one-on-one communication is the most trusted and that the farther up the educational hierarchy an individual gets, the less trusted they become. Certainly, these data support the findings detailed in Chapter 1 that the school superintendent and school board members should maintain a low profile. The finding also shows the need for using techniques that emphasize personal communications and that parents, teachers, and principals can deliver the message most effectively.

The media source most utilized by individuals is the newspaper, and the wise public relations committee will use local newspapers extensively. Gaining the support of the editor of the newspaper probably is critical to the success of the campaign. The committee can develop good relationships with the press by following a few simple suggestions[14]:

1. The person responsible for dealing with the press should respond promptly to any inquiry from the press.
2. If news media personnel can't make it to cover a story, the press relations chair should drop some notes off at the newspaper office.
3. Deadlines are a reality to newspaper reporters. The press relations chairperson should know when a story has to be in to make a particular issue or the advertising deadlines for a particular issue.
4. Complex or large numbers and figures should be rounded off.
5. A reporter should never be asked to show anyone a story before it is published.
6. Nothing is ever really "off the record."

If the committee does not have the financial and human resources to try to influence every voter, they should concentrate on those who are likely to deliver a "yes" vote. The "persuasion continuum"[15] provides a basis for this approach (see Figure 5.1).

The realistic goal of any type of persuasion is to move a person from where he/she presently is on the continuum toward a position closer to agreement. Those who are to the right of the undecided line will probably vote "yes"; those to the left of undecided will probably vote "no." While some individuals can be moved a considerable distance along the continuum, the best information and persuasive techniques in the world will move most people only a short distance along the line. Only those who start near the mid-point will move far enough to the right to turn their "no" vote into a "yes" vote. It is on those individuals that the public relations campaign should focus its efforts.

School bond issues can divide a community or they can bring groups within a community closer together. The public relations committee must make every attempt to insure the latter occurs. They must continually remember that there will always be another bond election.

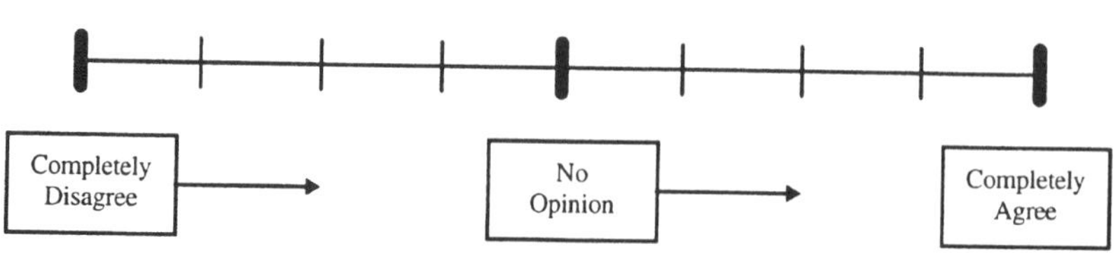

Figure 5.1. *Persuasion continuum.*

MARKETING TOOLS

In Chapter 6 several specific activities for school bond campaigns are detailed. Within those activities one can use several types of marketing tools. The following can be used for large meetings, all-community canvassing, small group activities, and media:

1. Flyers, brochures, and information sheets: Printed documents are important to any effort to persuade individuals. School boards can prepare and fund such documents as long as they do not promote a "yes" vote, but provide factual information only. Many boards of education send at least a brochure to all postal patrons in the district outlining the specific building project, providing information about the costs of the project and potential tax increases, and providing election information. This brochure is generally sent to all postal patrons within the district. While it is expensive, it is probably wise to have these brochures commercially produced.

   School boards can use flyers to announce informational meetings (tip: grocery stores might be willing to drop a flyer of this type in customers' grocery sacks). Flyers also can be tacked to posts in various parts of the community. This type of flyer is probably best produced in-house, using one of the desktop publishing programs now available for the computer.

   Information sheets should be distributed at large-group meetings, shopping centers or other areas where large groups of people gather, during door-to-door canvassing, or at small-group meetings. Again, the school board can pay for these sheets as long as they are informational, not promotional in content. These also may be most efficiently produced using desktop publishing.

   The campaign's public relations committee may also choose to use printed materials for its campaign. If such materials are produced with a promotional message, the committee should be sure to include a disclaimer on the document that states that it was paid for by the public relations committee—not the school board. The public relations committee may use the school board's documents if they so choose.
2. Video presentations: The combination of video and sound used in these presentations has proven to be effective in almost any type of marketing, and school bond elections are no different. In the past, the cost of such

programs has limited their use to public relations committees in larger communities where the committees had more funding. New, user-friendly technologies have, however, made it possible for many individuals to produce such presentations.

Video presentations offer several advantages. First, the committee can actually show district patrons the problems. They can show overcrowded classrooms, unsafe conditions, or structural concerns. Second, using video presentations can help assure that the same information is presented in the same way to all district patrons. Third, they are easy to use. Volunteers who might be reluctant to give a formal speech may feel comfortable putting in a video tape. Fourth, video presentations can be used at all sorts of meetings and gatherings—large or small. The chief disadvantage, as previously mentioned, is the cost.

3. Special events: Exposure is a key element in promoting any product, service, or (as in this case) idea. In most communities, there are a number of events or activities during which the public relations committee could have an informational/promotional booth where it could provide free information, sell products to raise funds (mugs, T-shirts, buttons), and conduct voter registration activities. These booths could also continually run the video presentation.
4. Media. Specific activities for utilizing the media are presented in Chapter 6; however, the importance of utilizing this resource cannot be overemphasized. Here, one must observe a note of caution. Board of education members and school officials should be careful about the amount of time they spend "in the spotlight." As noted in Chapter 1, it is critical that the school bond issue effort be perceived as a grass-roots endeavor, largely led by laypersons within the community. Constant appearances in the media by school officials do not promote that idea and should be avoided.

REVIEW ACTIVITIES

1. Describe how the features identified as important to improving the educational environment in our schools provide for better student learning, an improved quality of life for citizens in a community, and an improved business climate in the community.
2. Identify those factors in your "building projects" that some in your community might question and discuss the reasons they might object to those conditions.

3. Make a list of those in the media to contact, i.e., newspaper editors, television reporters, radio, and so on.
4. Summarize the reporting and editorial policies of all relevant media in your area.
5. Develop a media presentation for the public. The presentation should answer the following questions:
   a. What is the problem?
   b. What are the choices for fixing the problem?
   c. What are the recommendations from the committee(s)?
   d. How much is it going to cost?
   e. Where and when do we vote?
6. Analyze the "persuasion continuum."

ENDNOTES

1 Earthman, G. I. 1992. *Planning Educational Facilities for the Next Century*. Reston, VA: Association of School Business Officials International.

2 Castoldi, B. 1992. *Educational Facilities: Planning, Modernization, and Management*. Fourth edition. Boston: Allyn & Bacon.

3 Earthman, G. I. 1994. "How to Blow a School Bond Election—Or Not, If You'd Rather." *School Business Affairs*, 60(1): 3–8.

4 Marx, G., D. Bagin, and D. Ferguson. 1985. *Public Relations for Administrators*. Arlington, VA: American Association of School Administrators.

5 Earthman, G. I. "How to Blow a School Bond Election—Or Not, If You'd Rather." *School Business Affairs* 60(1): 3–8.

6 United States Government Printing Office. *U. S. Statistical Atlas, 1993*.

7 Castoldi, B. 1992. *Educational Facilities: Planning, Modernization, and Management*. Fourth edition. Boston: Allyn & Bacon.

8 ———.

9 Earthman, G. I. 1994. "How to Blow a School Bond Election—Or Not, If You'd Rather." *School Business Affairs* 60(1): 3–8.

10 ———. 3–8.

11 ———. 3–8.

12 ———. 3–8.

13 Marx, G., D. Bagin, and D. Ferguson. 1985. *Public Relations for Administrators*. Arlington, VA: American Association of School Administrators.

14 National Education Association. 1987. *Guide to Developing Positive Relations with the Press*. Washington, DC.

15 Watson, N. 1981. *Principles of Communication*. Boston: Allyn & Bacon.

CHAPTER 6

Campaign Activities

THE campaign committee can organize and implement several types of activities to achieve passage of the bond issue. The kinds of efforts they make will depend on the size of the community, the amount of money the committee has to spend, and the number of volunteers that have come forward. No matter what activities the committee decides to use, however, they should base them on a philosophy of openness and respect for the voter.

After conducting activities described in Chapter 2 of this book, the public relations committee should have a sizable list from which to solicit volunteer help. The committee should make every effort to seek volunteers from a broad cross section of the community. When seeking volunteers, the committee should be certain that those asked have a choice as to those activities in which they will participate. Volunteers should be given the following information about each activity: the type of tasks involved in each activity, the skills that might be required to do the tasks, the amount of time each task will require (estimated), and whether there would be any monetary obligation on the part of the volunteer. It is also important to remember that for some people, volunteering two hours after work is as much of a sacrifice as volunteering two days a week is for others. Be sure that the volunteers feel comfortable doing the work, and that they understand how valuable any contribution is to the campaign.

The number of activities the committee can plan will depend on how much money they can anticipate having to run the campaign. Most people do not like to ask for money from anyone, but to be successful the committee will need to purchase some services, pay for some commercial printing, and fund postage and office supplies. The total can reach into the thousands in very large communities to a few hundred dollars in small communities. The committee cannot depend on the school district to provide any funding for promotional activities, in that most states have laws prohibiting such contributions by a government agency.

When beginning the process of fund-raising, the committee should consider obtaining both large and small contributions. Small contributions will come from individuals who support the bond issue and are willing to give $5.00 to $10.00 for the effort. These funds can be raised with button sales, requests for donations at meetings or at informational tables, or through a fund-raising letter. Larger donations can be secured from individuals and businesses. Those businesses that would tend to profit from the project should be a first target.

Most individuals have had some experience running fund-raising events, so little space will be devoted to those activities in this chapter. The committee should make up an estimate of how much money they will need prior to initiating activities, so they know how much they have to raise.

Promotional activities should be aimed at those selected as a marketing target group. In Chapter 5, the discussion centered around those who are undecided but leaning toward voting "yes" and those who are undecided but leaning toward "no." These groups also tend to respond more positively to one-on-one contacts and word-of-mouth information. Therefore, the public relations committee should plan at least one personal contact activity.

ONE-ON-ONE CONTACT ACTIVITIES

Activity: Door-to-Door Canvassing

Purpose

1. To make a personal appeal to individual voters to vote "yes" in the election
2. To provide information to those citizens who remain uninformed about the election
3. To conduct voter registration and voting information
4. To gain a rough count of the potential number of "yes" votes

Number of Volunteers Needed

Approximately 1–2 per three-block area.

Tasks

The volunteers walk door-to-door in a specific neighborhood. At each house, they ask the occupants if they favor the bond issue. If they have decided to vote "no," the canvasser should thank them for their time and wish them a good day. The volunteer should not engage them in debate. If the occupants are undecided, the volunteer should ask them if they have any specific questions. If they do, the volunteer should answer as completely as possible or direct them toward a more authoritative source. The volunteer should leave a promotional brochure and/or information sheet at each house in which the voter is undecided. If the occupants answer that they are in favor of the issue, the volunteer should thank them for their support, provide them with a question and answer sheet (if any of their friends or neighbors need more information), and make sure they are registered to vote and know when the election is. The volunteer may also determine if the voter would need assistance getting to the polls on the day of the election. The canvasser should be sure to note the name, address, and telephone number of all "yes" voters, so that other volunteers can follow up later.

Timing of Activity

The canvas should be conducted early enough so that new voter registrations meet the deadline.

Materials

Promotional brochures, information sheets, list of addresses for appropriate marking.

Other Requirements

The volunteers should have a minimum of one hour of training during which they are taught interpersonal communication skills, review the materials they will be distributing, and are taught how to record the information they are to obtain.

Costs

Costs will vary according to the size of the community. The only major costs are those for printing materials. The brochure should be professionally printed; the information sheet can be a computer generated document.

Volunteer Commitment

Two hours for training and three to five hours for canvassing (depending on the size of the neighborhood).

Advantages

Allows the person-to-person contact that most people prefer, is relatively inexpensive, provides an opportunity to get good data on the number of "yes" voters. This activity works well in all communities.

Disadvantages

Requires a large number of volunteers.

Activity: Telephone Canvassing

Purpose

1. To gather information about the number of "yes" voters
2. To disseminate information to any individual who is still uninformed about the election
3. To provide information about voting on election day and determine if some of the voters will need any assistance getting to the polls (if they are "yes" voters)
4. Good back-up activity for door-to-door canvassing

Number of Volunteers Needed

Approximately one for every 50–70 voters.

Tasks

The volunteer receives a list of from 50–70 registered voters (voter list can be obtained from county offices, but they are a little expensive). Usually, those who have already indicated they will be voting "no" should be crossed off the list, if that is at all possible. The volunteer telephones the individuals on the list and asks them if they are in favor of the proposed bond issue. If they say they are going to vote "no," the volunteer thanks them for their time and wishes them a good morning (afternoon or night). If they are undecided, the volunteer should ask if they have any questions that the volunteer could answer. Volunteers can also ask if individuals need an information sheet. The volunteer should remind persons who indicate they are in favor of the election date and also determine which individuals need assistance getting to the polls. Some people do not like giving out any kind of information about how they are going to vote. If that is the case, the volunteer should thank them for their time and wish them a good morning/afternoon/evening. No calls should be made before 9:00 A.M. or after 9:00 P.M.

Timing of Activity

Usually, one week before the election.

Materials

Telephone script, voter registration lists with telephone numbers, forms on which to record information needed for follow-up activities.

Other Requirements

The volunteers should receive at least one hour of training on how to conduct canvassing activities. They also should have time to practice some phone calls with a helper present.

Volunteers' Commitment

One hour of training and about five hours of calling time. The time could be spread out over a few days. They also must have their own telephone.

Costs

Costs of copies of script, voter registration lists, and copies of forms.

Advantages

Inexpensive way of reaching many voters, provides one-on-one contact, good way for volunteers to get involved with smaller time commitment.

Disadvantages

Requires many volunteers.

Activity: Coffee Parties or Small Meetings

Purpose

1. To provide information to undecided voters
2. To provide information on voting in the election
3. To solicit volunteers
4. To raise funds

Volunteers Needed

This will vary considerably. Depends on the willingness of individuals to sponsor the activity.

Tasks

The host for the event invites from seven to fifteen friends or acquaintances to a coffee in the host's home to hear a presentation by a member of the public relations committee, a teacher, principal, etc. about the school bond election (those individuals who are trusted most, see Chapter 5). The host provides refreshments. The event can be formally organized (i.e., the speaker gives a presentation followed by a question/answer period before lunch) or informal (the speaker talks to each person in attendance on an informal basis during refreshments). The host should also provide some type of space to display informational and promotional materials.

Timing of Activity

Usually at least one month prior to the election. If voter registration is a goal, it should be timed so that the registrations can meet the deadlines.

Materials

Informational and promotional brochures and sheets. Voter registration forms and volunteer cards.

Volunteer Commitment

The volunteer will probably spend at least two hours in preparation, two to three hours for the event, and two hours on cleanup. In some cases, the cost of providing coffee and food for the guests would represent a financial burden to the host, so the committee should volunteer to pay for these. Most volunteers will probably refuse the offer, but it makes it easier for those who need to accept.

Costs

The promotional brochures, information sheets, and promotional items are the same as those used in other efforts, so the only new cost would be for refreshments.

Advantages

The informal atmosphere is conducive to participation. The setting allows many people, who would not speak at or even attend a large group meeting, to air their views and questions. This type of activity works particularly well in minority and lower socioeconomic segments of the community.

Disadvantages

The time commitment for volunteers.

MEDIA

Use of both print and electronic media can be critical to the success of any campaign. The following activities can help the public relations committee make the best use of mass media resources.

Newspaper Letters to the Editor

Purpose

1. To disseminate information to the community in a non-threatening manner
2. To advance the reasons people should vote "yes" to the proposal
3. To keep the issue in front of the public
4. To counter opposition viewpoints

Number of Volunteers Needed

The numbers will vary according to the size of the community and the number of newspapers. In smaller communities, the committee should pay attention to weekly papers that some segments of the community might receive. In larger communities, the committee should pay attention to ethnic, local community, and specialty newspapers. Volunteers for this task should have some writing training, or the committee should provide someone to proofread letters before they are sent to the paper. The committee would find it advantageous to utilize volunteers from as many segments of the community as possible.

Tasks

Volunteers should be recruited to monitor letters to the editor in local newspapers. This may be the only place that opposition views emerge, in that it is unusual for opposition groups to conduct an organized campaign in school bond elections. Those who review the papers then refer the topic to one of a list of volunteers he or she has to write a response. Care should be taken that the same individuals do not continuously write responses to letters. Specific individuals should be assigned dates to write letters to the editor on specific issues. For example, when the board chooses the school design, a volunteer might write a letter praising the design and indicating what advantages it has over other designs.

Timing of Activity

This effort is ongoing from the beginning of the investigation phase of the campaign. It should continue until election day.

Materials Needed

The volunteers should have access to appropriate information and resources.

Volunteer Commitment

The amount of time each volunteer devotes to this task is variable. In some communities, the volunteer will have several letters to write, while in others a volunteer may be asked to write only one or two letters.

Cost

The cost of postage to send letters. In some instances, the committee may wish to supply paper and envelopes.

Advantages

Provides an organized way of dealing with opposition communications so that everyone is certain a response will appear. It is inexpensive and can provide valuable clues as to what aspects of the proposal are receiving the most opposition. It also gives the PR committee and administration ideas on how to develop a plan for dealing with a specific issue.

Disadvantages

Care must be taken to avoid engaging in paper wars.

Newspaper Articles

While most newspapers have reporters specifically assigned to cover educational issues, those individuals cannot be expected to know about everything that is going on. No reporter turns down the opportunity to get a "tip" or information about a specific activity.

Purpose

1. To disseminate information about the proposal and school bond election
2. To highlight the type of activities the new facility would accommodate
3. To provide a response to opponents of the bond issue

Number of Volunteers Needed

Variable, depending on the number of articles, etc.

Tasks

Persons volunteering for these types of activities could participate in various tasks:

1. Notification of news reporters of special or newsworthy events
2. Sharing expertise in certain areas for interview purposes: For example, if the bond issue includes new facilities for assisting disabled students in home economics, the county extension agent or home living counselor might do an interview on the special problems faced by the disabled in managing their day-to-day lives.
3. Sharing expertise about specific concerns: For example, if part of the new facility is being arranged for Internet access, some parents may be concerned about their children locating pornographic sites. A person in the field might want to do an article on various types of security measures individuals and organizations can take.

Timing of Activity

This activity will be ongoing during the campaign as needed.

Materials Needed

Not a consideration.

Volunteer Commitment

Variable, but most can expect to spend from two to four hours for preparation and reporting.

Cost

None.

Advantages

Many people read the newspaper, it is inexpensive, and it is a non-threatening way of dealing with opposition points of view, and may involve school officials if all they do is disseminate information.

Disadvantages

The article is at the mercy of space, in some socioeconomic segments of the community few people read the paper, and there is always the problem of a reporter misinterpreting certain statements or misquoting the interviewee.

Newspaper Advertisements

Purpose

1. To provide announcements about special events and meetings
2. To promote voting "yes" on the bond issue

Number of Volunteers

Should be handled by the public relations committee only.

Tasks

Determine the type of advertisement that should be run and design and write copy for the ad. Generally, advertisements are used sparingly (the paper might run a "free" article). They should be used to announce important general meetings.

Timing of Activity

Only when needed during the campaign.

Materials Needed

No specific materials are required. In many cases, people in the advertising department of the newspaper are assigned to help write and design the advertisement.

Volunteer Commitment

This type of activity should not take a volunteer more than a couple of hours to complete.

Cost

Most advertisements are sold by the column length. Price depends largely on the rates that newspaper charges and that is quite variable.

Advantages

Good way of providing very specific information to the general public, many people pay more attention to ads than to full-length articles.

Disadvantages

Cost is high.

Radio/TV

Electronic media reach hundreds of people and in larger cities are an important resource. Although advertising on these media is usually cost prohibitive for most bond referendum campaigns, radio and television stations do offer some possibilities for free publicity. Most of these opportunities, however, require an objective viewpoint and should be thought of largely as opportunities for disseminating information.

Public Service Announcements and Event Calendars

Most radio and television stations reserve time each day to present an area calendar of events. The PR committee should make sure they understand the deadlines for having meetings announced (usually it is more than two weeks prior to the event), what types of information need to be provided, and any restrictions.

Community Talk Shows

One way of utilizing the media is to suggest appearances by individuals involved in the bond issue for the community talk shows that most stations run. This would be particularly important during the time when the building proposal is being announced. Such shows offer an opportunity for school officials to explain the issue, much in the same way they would during the large introductory meeting. Most stations are looking for material of interest to the entire community.

Some stations run point-counterpoint types of programs in which both sides of an issue are debated. If there is considerable controversy and if a representative from the opposition is found who is willing to appear on television, this may be an opportunity worth investigating. It should be noted, however, that the person representing the proponents of the bond issue may not be school officials, teachers, or others associated with the school. One needs to be cautious about appearing on such programs, and only persons who are familiar dealing with opposition views and conflict management should attempt such a project.

Direct Mail

Purpose

The public relations committee needs to reach as many voters as possible and should consider using at least one direct mailing to all patrons of the district. This mailing should not be confused with direct mailings done by the board of education, which must be objective presentations of the facts. This mailing is a direct appeal to the voter to vote "yes," provides reasons for voting "yes," and provides information on where and when to vote.

Number of Volunteers

This number will vary according to the size of the district and the method the public relations committee chooses to use for processing the mailing.

Tasks

At least one person will be in charge of designing the brochure and preparing it for printing. Whether you need many more volunteers will depend on whether the committee chooses to hire a bulk mailing service or decides to purchase its own bulk mailing permit and prepare the mailing. Such an endeavor requires many volunteers.

Bulk mailing services will attach labels, sort the mailings by zip code, and will use their own bulk mailing permit. The committee will need volunteers to deliver and pick up the brochures from the service and to deliver the items to the post office. In small school districts, it is probably about the same cost to use a service as to purchase the mailing permit. If the committee chooses to prepare its own mailing, they must see that the above tasks are done correctly.

Timing of Activity

The mailing should be made at least two weeks prior to the election. One must add time because mail handlers deal with first, second, and third class mail before handling bulk mail and the brochures could sit for awhile before the post office gets to them.

Materials Needed

Depends on the process one uses.

Cost

Bulk mailings are relatively expensive in that the committee must pay for printing the brochure and the mailing. However, such brochures can be effective and do reach most patrons of the district.

Advantages

The brochures reach a large number of people and deliver the message in a concise manner.

Disadvantages

The activity is relatively expensive compared to other activities and, if volunteers process the mailings, a large number of volunteers are needed.

REVIEW ACTIVITIES

1. Using the size of your school district as the determining factor, calculate the number of volunteers you would need to fulfill each type of campaign activity.
2. Determine the number of volunteers you would need based on the number of ethnic, social, and economic groups in your community.
3. Identify the organizations from which volunteers might be solicited in your community.
4. Make a list of special television and radio programming in your community for which the bond election campaign topic might be suitable.
5. Ascertain the costs of newspaper, radio, and television advertising in your community.
6. Research bulk mailing services in your community.

CHAPTER 7

Bond Election Day

NO day for a school superintendent or bond referendum campaign worker is filled with quite so many mixed emotions as election day. On the one hand, there is the excitement of the potential victory; on the other hand, there is the dread of a defeat. On the one hand, one wants to hide away in a closet until it is all over; on the other hand, and fortunately for stress levels, one has a myriad of tasks to accomplish.

POLL WATCHERS AND FOLLOW-UP ACTIVITIES

One must take a word of caution here. The general chairperson for the campaign should check the election rules of his or her state or school district carefully and inform all volunteers of state regulations limiting campaigning, location of signs and buttons, and which people may be present during certain election activities. Knowing the law can help avoid problems in the future in that the opposition will often call "foul" over a technicality in the election procedures if the bond issue passes. Most people are above such behaviors, but one should avoid the problem whenever possible.

On election day, the bond referendum committee needs to organize three sets of volunteers: poll watchers, telephone bank callers, and drivers. The committee chairperson for polling activities must have a considerable number of volunteers for the various activities. The key to winning any election is getting the voters who are going to vote "yes" to the polls. Every one of those votes counts, and the committee must make an effort to see that those votes are cast.

Poll watchers sit by the election officials and utilize their own voter registration list to mark off names as each voter comes through the line. The list is marked with how every voter responded to the poll about how they were going to vote. The committee should have enough poll watchers to change every one or two hours. As the workers change, the one going off duty reports to the campaign headquarters as to those individuals who previously reported they would vote "yes" who voted during their shift. This service is important for two reasons. First, the tally provides information as to how the election is going and what the anticipated results might be. Second, it gives the workers a list of potential "yes" voters who have not as yet made it to the polls.

The telephone bank callers can swing into action after the first few hours of balloting by calling the "yes" voters who have not voted as yet. These individuals should be encouraged to go to the polls and should be offered assistance, such as a ride, if they need it. Usually individuals respond favorably to these efforts. This activity is heightened during the last few hours of balloting. The callers may call some voters a second time to make a plea for them to vote. In a tightly contested election, these phone calls can make the difference.

By the end of balloting, the campaign committee should have a fairly good idea of how the election is going to turn out. They should not get overly optimistic, however, if the balloting seems to be going their way. In these types of elections, many people will tell telephone pollers that they are going to vote "yes" even if they are not because they fear being hassled over the phone. No one really knows what the outcome of the election will be until the votes are all counted.

CANVASSING BALLOTS AND REPORTING

After the polls close, election officials begin the process of counting ballots. This is a nerve-wracking time for supporters, board of education members, and the school superintendent. The best advice, however, is to stay away from the vote counting area. Election officials are under enough pressure to complete the process without having to contend with a large group of people milling around. Everyone knows it will take some time to tabulate all the results and gather them at one place. This might be a good time for a cup of coffee or antacid, depending on how each individual is feeling.

Once the officials complete the count, the school board meets to canvass the ballots and approve the election results. This procedure varies from state to state, and each school board will need to check what their procedures should be. The school superintendent is responsible for reporting the election results to the public and the news media.

WHEN THE RESULTS ARE UNFAVORABLE

Throughout this book, the authors have encouraged a positive attitude, but there is always the possibility that, no matter how hard everyone tried, the bond issue will fail. Even though most people who worked on the campaign will be disappointed, school administrators must maintain a positive attitude with the media and in public. It is probably sufficient to indicate that the committees will be meeting to review the results and determine the best action for the future.

The support committees should meet within the next week to determine what course of action they should take. Undoubtedly, most of the members of the committees will have discussed what opposition ideas defeated the election: if the result was because the issue was asking for too much money, or whether other factors entered into the election. The committee should then decide how they are going to mount the next campaign. The architect will probably have ideas as to how to lower the cost of the building or what design modifications might make the building more appealing. Most bond consultants will review the numbers to determine whether other funding alternatives might be available to lower the tax levies the bond issue would create. An analysis of which segments of the community voted in certain ways also can provide clues as to where efforts need to be improved.

The problem that brought about the bond election has not gone away. Another bond election will be held.

WHEN VICTORY HAS BEEN ACHIEVED

Sometimes, people have a difficult time controlling their excitement, and certainly a victory in a school bond election is one of those situations in which one may find it difficult to remain dignified. However, it is important for school officials to remain calm and appear objective in reporting positive results. One must be careful not to make any comments that might be construed by the opposition as criticism of them or their views. The school superintendent should be particularly cautious about answering any questions from the media about his or her feelings about the opposition. Officials must continually remember that other elections will be slated in the future.

One should not, however, try to calm the celebration of those who worked long hours to achieve victory. It is important that the school superintendent and board of education members maintain a low profile. It is sufficient for these individuals to stop in at celebration parties and thank those who worked to bring about the victory.

THE REAL CHALLENGE

After victory has been achieved at the polls, the real challenge of building the new facility or overseeing a renovation begins. This process is not a matter of just watching the walls go up; the school superintendent and board of education will face difficult dilemmas as changes need to be made and controversies arise.

An important occasion is the groundbreaking ceremony. Those invited to "make the first hole" should include school board members, representatives of the faculty and support staff, student representatives, community leaders, leaders of the bond election campaign, city and state officials, the architect, and contractor. Elaborate speeches do not have to be part of the ceremony, but it is always fitting for the school superintendent to address the public. If the building is to be named after an individual or group who was a member of the community, that person and/or members of his or her family should also be invited to participate.

While the building is in progress, the school superintendent should make every effort to keep the public informed as to the progress of the building, whether the facility will be finished by the estimated deadline, and any cost overruns he or she anticipates. Providing this information is probably handled best through newsletters or through reports in the local papers or electronic media. Any perception on the part of the public that information is being withheld from them or business is being conducted in secret can spell disaster for future bond elections. The public generally does not like surprises.

The superintendent should arrange times with the construction engineer and/or architect when it would be convenient and safe to conduct mini-tours of the facility as it is under construction. The school superintendent should invite persons from as many segments of the community, including representatives of groups that opposed the bond issue, to participate in these tours.

The fun part of the building process is planning the dedication ceremonies and open house when the building is completed. How elaborate these festivities are is a matter of personal choice. Most of these ceremonies include speeches by the school superintendent, the president of the board of education, and community leaders. Persons who participate in the open house should be provided with maps of the facility to make their unguided tour easier. Some school districts have their teachers in the classrooms, so that they can point out unique features in their rooms or improvements for the students.

When the hoopla is over, it is time for the school superintendent and the board of education to return to their five-year facilities plan and begin the process for the second project on the list (see Chapter 2). Providing safe and adequate facilities in which students may take advantage of the best educational opportunities possible is a never ending task. In the end, however, most find the prize well worth the effort.

REVIEW ACTIVITIES

1. Based on the number of precincts in your district, determine how many volunteers will be needed as poll watchers, telephone bank callers, and so on.
2. Research the election laws in your state and district to determine election-day procedures.

CHAPTER 8

Developing Special Skills

DURING any election campaign, individuals may be asked to assume roles for which they are not accustomed. Such roles include conducting surveys, dealing with conflict, and writing formal reports and other documents. Filling these roles successfully often demands the use of special skills. It is beyond the scope of this book to completely detail the theory and process of interpersonal and written communication. However, school bond campaign leaders may find the following suggestions helpful in all stages of the campaign.

The chapter includes three sections: conducting surveys, managing conflict, and preparing written documents. Each section will include some basic principles that underlie that section and suggestions for implementing those principles.

CONDUCTING SURVEYS

It probably is not possible or even advisable to conduct truly "scientific" surveys for school bond election referendums. Besides the expense involved, scientific data is not really necessary for activities outlined in this book. However, the goal should be to get as objective data as possible.

Both the board of education and the public relations committee for the referendum campaign can use several types of surveys: person-to-person, telephone, and forms. Each school district is unique in terms of its needs, and the public relations committee must determine which type of survey best fulfills its needs. The following suggestions for conducting surveys apply to all three methods:

- The sponsoring organization must clearly define the purpose(s) of the survey.
  The committee should determine what information it wants and/or needs to obtain from the survey. They should be cautious to limit the purposes to no more than three. If there are too many types of information, the survey gets too long and studies have shown that accuracy of comments goes down the longer the survey goes on. The committee should determine what its priorities are at the time. For example, the school board wants to obtain the following information from a preliminary survey (one conducted by the board before the proposal is organized): what type of support there is in the community for a school bond election, what priorities citizens in the community would have in a new facility, and the attitude toward education in the community. The survey at this point should not last longer than five to seven minutes. Obtaining all of the above information probably would take longer than that. Therefore, the board needs to choose the two pieces of information it considers the most important. If the new facility is being built to alleviate overcrowding, they would probably choose what type of support and attitudes toward education. If the building is being built because it is outdated in terms of the educational program, the board would probably choose what priorities patrons of the district would have and their attitudes toward education.
- Even though the organization does not have to use a complicated random sample formula, every effort should be made to contact various segments of the community. If the group is conducting a person-to-person survey, surveyors should be sent to the various residential areas of the district to conduct the survey, or they should be sent to shopping centers or special events located in different sections of the district. If the organization is conducting a telephone survey, they should select phone numbers from the various sections of the district. While not scientific, such selections do make getting an objective idea of what the community's attitudes are much more likely.
- Surveys must be non-threatening. There should be no indication that there is a right or wrong answer to any of the questions. People basically like to please other people, and, therefore, they will sometimes give the response they think is expected rather than an honest one unless they are encouraged to be forthright.

- Surveys should include questions that will check for reliability. For example, if one of the purposes of the survey is to determine what the major objections to the new facility are, at least two questions that should get the same or similar responses should be asked. One question might ask what patrons believe is the weakest part of the proposal. If some patrons are against the issue because of a raise in taxes, they will probably indicate that it is too expensive. The second question might be more direct, such as the patrons being asked to describe their major reason for objecting to the proposal. Variances in these two answers would indicate that something more than property tax rates is the reason for opposition.
- All persons who conduct the surveys should ask the same questions in the same way. The sponsoring organization needs to develop a script that all surveyors follow. The surveyors should not add their own comments nor should they make statements referring to a patron's response to a question.

Preparing the Survey Script

The script, then, becomes one of the most important elements of obtaining objective results. Certain elements are important to include:

1. The surveyors should identify the following at the beginning of conversations:
   a. Their names and the name of the organization for which they are conducting the survey
   b. The purpose of the survey and how the results are going to be used
   c. Who will have access to survey results (confidentiality)
   d. About how long the survey will take

   A good opening could be:

   "Good evening, Mr. ______. I'm ______ and I'm conducting a survey for the ______ district school board about the possibility of building a new ______ school. I was wondering if you had about five minutes to answer some questions so the school board has an idea of how the patrons of this district would feel about the issue."

If the respondents refuse, the interviewer should thank them for their time and wish them a good evening. The surveyor should never try to convince a patron to participate.

2. The questions should be as non-threatening as possible. Many surveyors recommend using contingency questions rather than direct questions. For example, when asking how the person is going to vote, the question might be phrased: "If the election were held tomorrow, how would you be likely to vote, yes or no?" Phrasing the question in this way allows the patron to answer in a less direct manner than if asked whether he or she is going to vote yes or no on the school bond issue. A typical response might be: "Well, I don't know much about it yet, so tomorrow I'd probably vote no." They don't have to come right out and say they're opposed to the bond issue. As the campaign nears completion, the survey questions may become more pointed, but for most purposes, contingency questions are a good approach.

   Using questions where the patron chooses from a list of responses is also effective.

"If you had to describe the reason you would vote "no" on the school bond issue, it would be:

a. It is too expensive.
b. We really don't need a new school.
c. I don't like the location they are talking about.
d. They've included too many frills."

One pitfall of this type of question is the inclusion of too many options. Sometimes the patron will become confused by many choices. Therefore, the number of choices should be limited to four or less. Ask another question rather than adding ten choices.

Another type of question is the open-ended question. In this type of question, the interviewer begins a statement and asks the respondent to finish it. For example, the interviewer might say:

"If I could tell the school board the two most important things I think should be in a school building, they would be: ______ ."

A pitfall of this type of question is getting vague answers or a lot of "I don't knows." Do not press respondents if they do not seem able to answer the question more precisely.

3. The surveyor should allow respondents to discontinue the interview at any time. Sometimes, patrons become uncomfortable answering questions or get bored (a good reason for keeping it short). The interviewer should not argue with the respondents, but should thank them for the responses they gave and say good evening.
4. At the end of the survey questions, the surveyor should thank the participant and assure him or her of how valuable the information will be.

An example of a good closing could be:

"That's all the questions I have, Mr. ______ . Thank you so much for taking the time to help me. The school board is trying to make a good decision about this project, and your input will be most helpful to them. Have a good evening."

Recording the Results

Every surveyor should be given a form on which to record the answers. This form should be easy to fill out. The slots for the name of the respondent, if it is being taken, should be at the top. If listing or contingency questions are used, the possible choices should be listed and all the surveyor has to do is circle the response given. Surveyors should add notes only if what the interviewee says leads to another point that fits into that category (e.g., the respondent giving a reason for opposition that is not on the list).

The form should have a survey or identification number on it.

Analyzing the Results

A separate group from those conducting the survey should be brought together to compile and analyze the results. This prevents persons from inadvertently adding an interpretation that might not be there. For example, a surveyor might say, "Well, I know he said he was going to vote 'no,' but I really felt like he wanted to vote 'yes'." The tabulators have no idea in what manner the questions were answered; they have only the answers.

Analysis of the results amounts to "counting up the responses." This is another advantage to offering specific choices to the participants. They are easy to count. This information can then be used to answer the questions the group is asking. For example, if the board wants to know what percent of the voters presently support the bond issue, the committee counts the number of "yes" responses and calculates the percentage of "yes" voters.

Responses to open-ended questions should be categorized and then counted. For example, if the committee is trying to determine the main reason for opposition and ask open-ended questions, they should group responses into categories such as money, building features, location, dislike of school district, and other. This eliminates trying to figure out what every response means. For example, the responses "Building costs too much," "Can't afford the taxes," and "Hate spending that much money on anything" all belong generally to the same category (money).

One should anticipate discrepancies in the survey results. For example, the number of "yes" votes is usually overreported in person-to-person surveys and telephone surveys because the respondent fears a confrontation. This means the interpretive committee needs to be cautious in reporting results so that no one gets too excited about the possibilities and counts on something that may not happen. The result could be a misinterpretation of why a bond issue failed.

MANAGING CONFLICT

In any situation in which there are opposing views, conflict will occur. Humans tend to avoid conflict whenever possible; however, there are specific techniques that can be used to manage conflict situations and avoid the negative results of adversarial relationships.

Communication Principles

Conflict management techniques are based on the following principles of interpersonal communication:

1. Effective interpersonal communication can exist only in an atmosphere in which both parties have a sense on equality of status. For example, in school bond elections, antagonism can begin if those in opposition believe their reasons for opposing the proposal are being dismissed by those favoring the issue as excuses for "stinginess" or as indications that individuals simply do not care about providing a quality education for the young people of the school district. In some instances resentments can build if the administrators, school board members, and campaign leaders are all perceived as being of a "higher" socioeconomic status than the general citizenry. Many times those in lower socioeconomic segments of the community feel that the "rich" people don't care about how many frills are included in the proposal because the amount of the raise in taxes does not represent as large a percent of their income as for those in lower income brackets.
2. Effective interpersonal communication occurs when individuals from both sides of a particular issue believe they are working toward a common goal. Every argument has some point at which all parties agree. For example, two parties can agree that children in our society need to be protected; they may disagree on who is responsible for insuring children's rights. The same is true for school bond election campaigns. The point of agreement may be basic: all children should get some education, but it is a point from which both sides can engage in discussion.
3. Effective interpersonal communication deals with ideas and behaviors—not with personalities. When placed in a confrontational situation, many people become defensive. They tend to strike back at their opponent rather than taking the time to find the flaws in their argument. For example, in many school bond issue campaigns, the opposition will discuss the "frills" of a gymnasium or theater. Sometimes those in charge use emotional appeals such as "we need to give our students the best" (inferring that those opposed do not want to do that). To say the least, these types of comments are off-putting to most individuals who are questioning the expense. They want some rationale—some good reason—for the inclusion of such facilities.
4. Effective interpersonal communication can occur only when both parties reveal their true ideas. In many instances persons who oppose school bond issues have so-called "hidden agendas." They are either

afraid or embarrassed to give the real reason for opposition. For example, if the location of a new facility would require children to drive or ride through what is known as a high crime area of a city, parents might be concerned about their children's safety, but they may hesitate to say anything because the area is populated mostly by one or two ethnic groups, and the parents do not wish to be considered racist. Antagonism grows (mostly on the part of individuals conducting the campaign) when individuals answer the arguments about taxes, which those in opposition are focusing on, and the opposition is still resistant. The campaign leaders simply may not be answering the real questions.

Communication Techniques

Many specific techniques can be used to improve interpersonal communication and, therefore, lessen conflict:

1. Prepare carefully. Part of the preparation process as noted in the previous chapters is identifying possible opposition points of view and individuals or groups. In terms of dealing with conflict situations, one should try to understand what the particular needs of those individuals and/or groups are and what types of information or actions would help fulfill their needs. For example, some individuals are fearful that a raise in property taxes would mean they could no longer stay in their homes. Providing these individuals with information about property tax assistance available to them could relieve some of their anxiety about their financial futures. Other individuals may feel that their children's safety might be compromised by their having to travel through unsafe neighborhoods to get to the new school. They need to see alternative travel routes, or perhaps the district needs to develop new school bus security measures that could be implemented to make them understand that their child's safety is important to the district.
2. When presenting information or trying to persuade, individuals should focus on points of agreement rather than on points of disagreement. If one can get a hostile individual to agree on some point, they can sometimes be moved to accept certain propositions. For example, if the individual is questioning the need for certain building features, such as a new gymnasium, the campaigner should seek that person's agreement on the children's need to be physically fit and/or the bene-

fits of extra-curricular activities. Getting agreement on any point can shift the argument to how people work together to accomplish that common goal.

3. Use contingency statements. Especially during the process of gathering information about attitudes about the schools and developing awareness of the need for the project, one should deal in terms of "if . . . then" statements. For example, if an individual has concerns about the safety of his/her child, one could offer the contingency statement: "If the school district implemented a new school bus security program, would you find the proposal more satisfactory?" Such statements give the person in opposition a chance to respond in a favorable way and provide important information to those running the campaign about the strength of opposition feelings. This also provides ways for people not having to "lock in" their position. Once people have been forced to do that, they often become stubborn about changing because they do not want to "lose face."
4. Rather than reacting immediately to statements made by hostile individuals, one should take time to "think" about their responses by using restating techniques. The meeting leader or other individual restates what he/she believes other individuals said: "What I hear you saying is that property taxes are already too high for many people to afford." Such restatements lengthen the time between the statement and the response, which sometimes prevents misstatements and "off-the-cuff" gaffs that can exacerbate an already tense situation. It also allows the other individual to clarify his/her statement if it has been misinterpreted.
5. One should respond in terms of how certain statements or actions personally affect the individual, rather than expressing a perception of the individual who made the statement. For example, if the person in opposition is talking about wasteful spending practices, a meeting leader might remark: "When you make statements about our wasting money, I become concerned because I haven't done a good enough job providing truthful information about our spending decisions to the public." Usually, people become less hostile toward individuals who express such concerns about their own behavior or feelings. Hostility would increase in the above situation if the comment was: "If you feel that way, you just don't understand all of the problems we have when deciding how to spend the district's meager funds."

6. Individuals should understand the importance of maintaining a sense of humor during tense situations. Humor tends to objectify the situation, relieve anxiety, and distract people away from the problem confronting them long enough for them to collect their thoughts.
7. Get to it first. If the campaign worker has the information, stating the opposition concerns and then answering their concerns before the opposition brings them up often has the effect of "taking the winds out of their sails." In other words, the proponents of the bond issue recognize the concerns of individuals and address them, without the opposition having to confront the proponents. This also gives opponents a better feeling because it indicates that the proponent groups have already considered their feelings.
8. Proponents should never try to leave a meeting or conversation angry. Taking a minute to shake the hand of an individual with whom one has had a confrontation keeps the issue from becoming a personal issue. One often hears the comment after tense meetings: "I guess I put that guy in his place." Taking a minute to acknowledge that the proponent disagrees on this issue, but does not dislike the individual with whom he/she disagrees might allow that individual to understand that two people can disagree about an issue without the other person being "bad." Such an impression allows for future communication.

These techniques are not easy to use. Most individuals need to learn these techniques. All campaign leaders and workers should receive training in such techniques and should be given an opportunity to practice using them before they deal with the public. This will give the workers more confidence in their ability to handle confrontational situations and help them from becoming defensive in hostile situations.

PREPARING WRITTEN DOCUMENTS

Many individuals fear writing more than any other activity. Whether it be writing a formal report, developing a fact sheet, or designing copy for a brochure or flyer, when confronted with a blank sheet of paper, many people panic. It is beyond the scope of this book to present a mini-course in writing, but the following suggestions should help the writer deal effectively with the first steps.

Keeping It Simple

The cardinal rule of writing is to use the simplest format, vocabulary, and sentence structure that will adequately convey meaning. Simplifying the structure allows the reader to proceed more quickly through the document, thereby improving comprehension. The writer should keep in mind the three basic parts of any document: the introduction, the explanation of major ideas and supporting materials, and the conclusion.

The introduction states the thesis of the report, identifies the purpose, and provides a basic overview of the report. An example of a good introduction for the site report might be:

"The site committee has examined three potential sites for location of the proposed elementary school. The committee examined each site as to topographic features, accessibility, suitability for proposed design, and geographic location. The committee then ranked each site in each category before coming to its final recommendation. The following report identifies various factors about each site that influenced the eventual recommendation."

The middle section provides those ideas that are being examined and provides supporting materials. When organizing this section, the writer can use a "briefs" method at first and can write it in complete sentences later. For example:

Topic Sentence: Site 1 provides easy access for most students in the area.
Support:

a. The distance from school to residential areas: Two blocks from south; 5 blocks from north; 6 blocks from east; one block from west
b. Risk factors: Students from the south, west, and east would not have to cross any major streets or highways. Students from the north would have to cross one major street, but stop lights are located every three blocks, improving the safety of students from the north.
c. And so on.

The conclusion is a brief summary of what the total report has stated.

For example:

The committee recommends that the board choose Site 1 as the location for the new elementary school project. The committee's recommendations are based on the site's high rankings in the categories of ease of access, site topography suitability, and availability of services. While more expensive to obtain than either Site 2 or Site 3, Site 1 would require fewer expenses in terms of construction on the site, providing safety guarantees and installing sewer and water, electricity, and telephone.

These three basic elements of a report or information sheet can be organized in several ways; however, two such formats provide easier writing techniques than does a straight narrative form.

Organizational Formats

When the writer is determining which type of organization to use, considering two formats may make the task easier: the outline format and the question/answer format. These two formats may be used for preparing reports, writing information and fact sheets, or designing flyers and brochures. They allow the writer to avoid some of the complications associated with writing in a narrative form, and the reader can follow the ideas easily.

Outline Format

This style uses the traditional outline as its basic structure. The writer may choose to use bullets or other means of identifying the different levels of information, but the principle is the same. The format is particularly useful when the writer is comparing two things (e.g., the report from the site committee), and when the writer is merely presenting data—not making a specific recommendation. For example:

Site Report
I. Ease of Transportation
  A. Site Option 1

1. Distance from residential areas.
   a. North: 15 blocks
   b. South: 5 blocks
   c. East: 10 blocks
   d. West: 4 blocks
2. Access.
   a. Car: The site is located along two major streets that provide through traffic easy access.
   b. Walking: Because of the busy streets nearby, students would have some difficulty walking to school. The situation could be improved by including walkways over the streets.
   c. Bicycle: Again, there would be difficulty for students trying to ride their bicycles to school.

The outline would proceed through the other criteria in that category and then through each site being considered. If several sites are being considered, the outline form might be adapted to place the information in columns.

Question/Answer Format

This format uses a question as the heading, followed by the answers in paragraph or outline form. This format allows great ease in comprehension because the attention of the reader is focused on the specific question. For example:

Question: Which site provides the greatest ease of access for students?

Answer: All three sites have advantages and disadvantages in this area; however, the committee believes that Site 2 provides the greatest safety and therefore is our recommendation.

Site 1 is located closer to residential areas in all four directions; however, its location close to two very busy streets would make it difficult for students to either walk or ride their bicycles to the site. We believe the students' safety would be compromised at this site unless over-the-street walk paths were included in the school design. Such structures would add considerably to the expense.

Site 3 is close to residential areas to the north and east, but the long distances from the west and south (30 blocks and 15 blocks) would mean considerable travel time for those students and would prohibit many of them from walking or riding their bicycles. Students from the west and south would also have to cross three major streets on their way to school, increasing the risk of accidents.

The location of Site 2 is central to residential areas in the four directions, with the distances students would have to travel being about equal (10 blocks, north; 6 blocks, east; 8 blocks, west; and 10 blocks, south). The major advantage of this site is that most students could reach the site by walking or bicycling without having to cross major highways or streets, thus increasing their safety.

This system also has a built-in check for writers. The writer can examine each answer to determine if it directly answers the question being posed. If the information does not, the author should delete it or ask a question that the information does answer. This check system helps the writer to stay organized and avoid the inclusion of extraneous information.

Choosing which format is best for your report or other document should be determined by its purpose and the writer's comfort using each format. The writer should also consider including highly technical material in an attachment rather than writing it in the main text. Technical information is usually jargon-heavy, and most readers will find it confusing. Putting lengthy technical discussions in an attachment allows the writer to keep the major portions of his or her document concise and to the point.

Vocabulary and Sentence Structure

The cardinal rule here is to analyze the needs of the audience. There are several points to consider when determining vocabulary and sentence structure.

1. Reading level: If the document is meant for a special committee that has experience in the area the writer is discussing, the reading level is not a consideration. However, if the document is meant to be read and understood by the general public, the writer should insure that the reading level does not exceed the eighth grade level. Most word processors now have systems that automatically calculate the reading level of any file.
2. Complexity: Here again, the writer should be concerned with "need

to know." If the writer is preparing an analysis of site topography for the architect, he or she will likely include most of the technical data. If, however, the document is intended to provide a justification for selecting a site to the general public, the writer needs to "get in, say what he or she has to say, and get out." The writer should not feel a need to provide all there is to know about a topic. In fact, in most business writing today, most experts advise including only essential data. The writer should be sure not to tell the reader "more than they want to know" just because he or she can.

3. Sentence types: The writer should use a variety of sentence types, but he or she should avoid using strings of lengthy sentences in any writing. Many writing experts believe the most effective writing includes approximately 40 percent simple sentences, 30 percent compound sentences, and 30 percent complex sentences. Certainly, it is difficult to have that kind of exacting control, but the writer should strive to avoid sentences that, because of their length, might prove difficult for the reader.

Getting Started

If there is any good advice for how to get started, it would be that the writer try not to do it by him or herself. Ask other individuals what type of information they think they would need, ask your committee or a trusted group to brainstorm the important concepts and then prioritize their concepts.

The writer also needs to organize all the supporting data before he or she begins to write. Different organizational schemes will work, but the most effective is to break the topic down into categories and organize information under each category together. This avoids the need to look through reams of paper to find a particular document or number. If the writer has the information organized in a logical system, he or she will save much frustration in the later stages of writing.

The writer should begin with a topic outline of some sort. Just a brief sketch of the ideas to be presented and the order in which they will be presented will provide the writer with a way to determine where more information is needed and what the next topic should be. The writer should have this basic outline handy at all times and should refer to it often to prevent getting "off track."

Tips for Proceeding

1. The writer should expect to revise. The first time through, the writer might put in question marks to himself or herself, indicating a missing word or other consideration. When the writer reaches a block (can't think of the right word or is unclear about what comes next), he or she should not ponder over it for hours before going on; such action only leads to frustration and wastes a great deal of time. "Put the problem away" until the next time through.
2. The writer should plan enough time for the project so that he or she does not have to work on it non-stop. Four to five hours is generally considered the maximum number of hours a day a writer can realistically spend in productive work. Time above and beyond that point is often wasted. Fatigue sets in, the mind slows down, and boredom begins to emerge. The writer should take breaks as often as needed. This is not advocacy of procrastination. The writer should set up a specific time each day to work on the report. Keeping a schedule is conducive to creativity and productive work.
3. The writer should let others read the work as he or she proceeds. Authors find editing their own work particularly difficult. Others often will catch errors that the writer misses. The writer needs to be able to take criticism and advice in the manner in which it is given—usually, the person who is offering advice is trying to keep the writer from looking stupid.
4. The writer needs to make use of dictionaries, thesauruses, books of quotations, grammar handbooks, and other reference materials as much as possible. If the writer is unfamiliar with how to use a particular reference book, he or she should consult a librarian or English instructor. Sometimes a writer has difficulty knowing when a reference book should be used. The adage to follow should be: when in doubt, look it up. All reference books should be kept close at hand, so that it isn't easier to guess than it is to look it up.

General Writing Suggestions

1. The writer should "get in, say what he or she is going to say, and get out." Avoid explanations of explanations, redundancies, and repeated words. Sometimes the writer finds it necessary to offer alternative ex-

planations of a concept because the concept is difficult to understand; however, redundancies when overused are confusing and boring for the reader. If an author uses the words "in other words" more than twice in an entire report, he or she should examine why the need for re-explaining. Examine why the first version is not sufficient to explain the process.

2. Also, the writer should avoid the use of jargon or very technical terms. The point of writing is to communicate ideas. If the reader does not understand the words being used, communication has not taken place.
3. Most writing students become confused by the amazing number of rules that govern formal writing. The beginning writer should not worry about "knowing all the rules" (e.g., subject/verb agreement, tense agreement, proper placement of prepositions, and so on). His or her best choice is to get the ideas and supporting materials down on paper and then find a proficient writer to edit for grammar. The beginning writer usually knows how to organize ideas, but becomes overwhelmed by the technical writing rules. No one should be ashamed to hand it over to an expert.
4. The writer should not hesitate to use a graph, table, or pictorial representation of an idea rather than write complete narrative. Sometimes, using such visual representations can save hundreds of words.

Writing clear, concise reports, information sheets, brochures or other documents is integral toward the organization of a successful election campaign. The above information is "bare bones"—some essential information. The conscientious writer may want to investigate other books on specific writing and designing tasks before he or she begins the process.

REVIEW ACTIVITIES

1. Design an appropriate survey to obtain preliminary information on community attitudes toward a school bond election. Then design a survey that campaign volunteers could use to determine attitudes toward the project about two weeks before the election.
2. List situations during a school bond election campaign that might

bring about conflict. Write a paragraph for each situation explaining how you would handle the problem.

3. Write practice portions of the report to the school board and have an editing specialist review your writing style.
4. Determine which parts of the total report to the board should use which type of organizational scheme.

State School Bond Referendum Percentages

State	Percentage	Statutory Notes
Alabama	0 or 50 + 1	**Title 39. Chapter 7, Improvement Authorities.** When a board of education decides to build a new building, the following procedures are required: (1) The board must select a site that meets the requirements of the state building committee; (2) The board must select an architect; (3) The architect must prepare the building plans and specifications and submit such plans at each stage of the process to the state building commission for approval—preliminaries, schematics, and final plans; and (4) The board must advertise the bids in a local newspaper for a number of weeks, in accordance with the state bid law. The state fire marshall must also approve the plans for the building. **Note**: If the board can convince the city council or the county commission to levy a sales tax, that can be done without a vote of the people. However, if a local property tax is to be increased, the local legislative delegation must submit a local bill to the state legislature for approval and the people must vote to increase the levy. If any part of the building is to be paid for by state building funds, the Alabama Public School and College Authority (PSCA) must approve all of the plans. (The procedures were provided by Dr. Harold Patterson, former school administrator in Alabama. Also, see Code of Alabama. 1975. Volume 20, Title 39, pp. 506-521.)
Alaska	50 + 1	**§ 29.47.190. Vote and notice of existing indebtedness required.** (a) A municipality may incur general obligation bond debt only after a bond authorization ordinance is approved by a majority vote at an election. Any municipal voter may vote in the bond election, except as otherwise provided by law. § **Sec. 14.11.100. State aid for costs of school construction debt.** (a) During each fiscal year, the state shall allocate to a municipality that is a school district, the following sums: (1) payments made by the municipality during the fiscal year two years earlier for the retirement of principal and interest on outstanding bonds, notes, or other indebtedness incurred before July 1, 1997, to pay cost of school construction, (2) 90 percent of (A) payments made by the municipality during the fiscal year two years earlier for the retirement of principal and interest on outstanding bonds, notes, or other indebtedness incurred after June 30, 1977, and before July 1, 1978, to pay cost of school construction. . . . (7) . . . 70 percent of the payments made by the municipality during the fiscal year for the retirement of principal and interest on outstanding bonds, notes, or other indebtedness authorized by the qualified voters of the municipality after March 31, 1990, but before April 30, 1993, to pay costs of school construction, additions to schools, and major rehabilitation projects. (Alaska Statutes. 1962. Volume 7, pp. 254-257.)
Arizona	50 + 1	**§ 15-1023. Issuance of bonds.** Upon receipt of the certificate of the appropriate election officer as provided in § 15-493 that a majority of votes cast at the bond election favors issuing the bonds, the school district governing board shall issue the bonds of the school district. (Arizona Revised Statutes. 1997. Volume 6A, p. 102.)

Arkansas	50 + 1	**§ 6-20-1205. Submission of statement prior to borrowing money or issuing bonds.** Approval by the State Board of Education or by the Director of the State Department of Education, General Education Division. (Arkansas Code. 1997. Supplement, Volume 4, p. 222.) **§ 3. School district tax—Budget—Approval of tax rate by electors.** The General Assembly shall provide for the support of common schools by general law, including an annual per capital tax of one dollar to be assessed on every male inhabitant of this state over the age of twenty-one years; and school districts are hereby authorized to levy by a vote of the qualified electors respectively thereof an annual tax for the maintenance of schools, the erection and equipment of school buildings and the retirement of existing indebtedness, the amount of such tax to be determined in the following manner: The Board of Directors of each school district shall prepare, approve and make public not less than sixty (60) days in advance of the annual school election a proposed budget of expenditures deemed necessary to provide for the foregoing purposes, together with a rate of tax levy sufficient to provide the funds therefor, including the rate under any continuing levy for the retirement of indebtedness. If a majority of the qualified voters in said school district voting in the annual school election shall approve the rate of tax so proposed by the Board of Directors, then the tax at the rate so approved shall be collected as provided by law. In the event a majority of said qualified electors voting in said annual school election shall disapprove the proposed rate of tax, then the tax shall be collected at the rate approved in the last preceding annual school election. (Arkansas Code. 1987. Constitutions, pp. 286-287.)
California	66⅔	**§ 15124. Entry upon favorable vote; certification of results.** If it appears from the certificate of election results that two-thirds of the votes cast on the proposition of issuing bonds of the district are in favor of issuing bonds; or, a majority of the votes cast, if the election is held to repair, reconstruct or replace school buildings in compliance with Section 17367 or 81162. (California Codes. 1998. Education Code. Volume 26A, p. 90.)
Colorado	50 + 1	**§ 22-43.5-110. Election—declaration of organization.** If a majority of the votes cast at the election are in favor of the capital improvement zone and the issuance of bonds and if the board of education determines the election was held in accordance with article 42 of this title and any other laws applicable to the election, it is authorized to proceed. (Colorado Revised Statutes. 1997. Volume 5A, p. 194.)
Connecticut	50 + 1	**§ 10-56. Corporate powers. Bond issues.** (a) A regional school district shall be a body politic and corporate with power to sue and be sued; to purchase, receive, hold and convey real and personal property for school purposes; and to build, equip, purchase, rent, maintain or expand schools. Such referendum shall be conducted in accordance with the procedure provided in section 10-47c except that any person entitled to vote under section 7-6 may vote and the question shall be determined by the majority of those persons voting in the regional school district as a whole. (Connecticut, General Statutes Annotated. 1958. Volume 5A, p. 136.)
Delaware	50 + 1	**§ 2123. Election result; certification.** The State Board of Education shall make out a certificate of the result of such vote which shall be filed and kept in the Offices of the State Board of Education as a public record. If at such election a majority of the vote cast throughout said district shall be for the bond issue, then bonds to the amount voted upon shall be issued as provided for by this chapter, but if at such election a majority of the votes cast shall be against the bond issue, then the bond issue proposed shall not be made. (Delaware Code Annotated. 1974. Volume 8, 1993 Replacement Volume, p. 388.)

Florida	50 + 1	**§ 236.41. Result of election held.** If it shall appear by the result of said election that a majority of the votes cast shall be "For bonds," the school board shall be authorized and required to issue the bonds authorized by said election for the purposes specified in the resolution as published, not to exceed the amount therein named; but, if the majority of the votes cast shall have been "Against bonds," no bonds shall be issued. (Florida Statutes Annotated. 1989. Volume 11A, pp. 163-164.)
Georgia	50 + 1	**§ 36-82-3. (GCA § 87-203) Bonds may be issued, when.** (a) When notice has been given and the election has been held, in accordance with Code Section 36-82-2, if the requisite majority of those qualified voters of the county, municipality, or political subdivision voting at the election vote for bonds, then the authority to issue the bonds in accordance with Article IX, Section V, Paragraph I or II of the Constitution of Georgia is given to the proper officers of the county, municipality, or political subdivision. (Code of Georgia Annotated, Book 45. 1994. Revision, p. 552.)
Hawaii	66⅔ (State Legislature)	**§ 39A-116. Issuance of special purpose revenue bonds to finance projects.** The legislature finds and determines that the exercise of powers vested in the [Department of Education] by this part constitutes assistance to a processing enterprise and that the issuance of special purpose revenue bonds to finance facilities of, or for, or to loan proceeds of such bonds to assist, project parties, is in the public interest. However, under **§ 39A-117, Authorization of special purpose revenue bonds,** special purpose revenue bonds for each project or muliproject program shall be authorized by a separate act of the legislature, by an affirmative vote of two-thirds of the members to which each house is entitled; provided that the legislature shall find that the issuance of such bonds is in the public interest. (Hawaii Revised Statutes Annotated. 1995. Volume 1, pp. 527-528.)
Idaho	66⅔	**Article VIII. Public Indebtedness and Subsidies. Section 3. Limitation on county and municipal indebtedness.** No county, city, board of education, or school district, or other subdivision of the state, shall incur any indebtedness, or liability, in any manner, or for any purpose, exceeding in that year, the income and revenue provided for it for such year, without the assent of two-thirds of the qualified electors thereof voting at an election to be held for that purpose, nor unless, before or at the time of incurring such indebtedness, provisions shall be made for the collection of an annual tax sufficient to pay the interest on such indebtedness as it falls due. (Idaho Code, General Laws of Idaho Annotated. 1993. Volume One, p. 300.)
Illinois	50 + 1	**§ 34-22.2 Issuance of bonds.** For the purpose of erecting, purchasing, or otherwise acquiring buildings suitable for school houses, erecting temporary school structures, erecting additions to, repairing, rehabilitating and replacing existing school buildings and temporary school structures, and furnishing and equipping school buildings and temporary school structures, and purchasing or otherwise acquiring and improving sites for such purposes the [school] board, with the consent of the city council expressed by ordinance, may incur an indebtedness and issue bonds therefor in an amount or amounts not to exceed in the aggregate $50,000,000 in addition to the bonds authorized under Section 34-22.1. These bonds, [not to exceed 20 years], shall not be issued until the question of authorizing such bonds has been submitted to the electors of the city constituting said school district at a regular scheduled election and approved by a majority of the electors voting upon that question. In addition to the requirements of the general election law the notice of the referendum shall contain the amount of the bond issue, maximum rate of interest and purpose for which issued. (Illinois Compiled Statutes Annotated. 1993. Education, p. 143.)

Indiana	------	**§ 6-1.1-20-2. Issue authorized.** — A political subdivision may, subject to the limitations provided by law, issue any bonds, notes, or warrants, or enter into any leases or obligations that it considers necessary. However, **§ 6-1.1-20-3.2. Property taxes to pay debt service or lease rental—Request for application of petition and remonstrance process—Procedures [as amended by P.L.2-1997].** — If a sufficient petition requesting the application of a petition and remonstrance process has been filed as set forth . . . of this chapter, a political subdivision may not impose property taxes to pay debt service or lease rentals. (6) If a greater number of owners of real property within the political subdivision sign a remonstrance than the number that signed a petition, the bonds petitioned for may not be issued or the lease petitioned for may not be entered into. (Indiana Statutes. 1988. Title 6, Articles 1-3.5, pp. 3898-395.)
Iowa	60 or 50 + 1	**§ 75.1. Bonds—election—vote required.** When a proposition to authorize an issuance of bonds by a county, township, school corporation, city or by any local board or commission, is submitted to the electors, such proposition shall not be deemed carried or adopted, anything in the statutes to the contrary notwithstanding, unless the vote in favor of such authorization is equal to at least sixty percent of the total vote cast for and against said proposition at said election. (Iowa Code Annotated. 1991. Volume 4A, p. 360.) **Section 1. <u>New Section</u>. 422E.1 Authorization—Rate of Tax—Use of Revenues.** House Bill 2282, passed and signed by the Governor in the 1998 legislative session, authorizes "the imposition of a local option sales and services tax and use of certain federal funds for school infrastructure projects and the issuance of bonds, and providing for an effective date. **2.** The maximum rate of tax shall be one percent. The tax shall be imposed without regard to any other local sales and services tax . . . and is repealed at the expiration of a period of ten years of imposition or a shorter period as provided in the ballot proposition. **Section 2. <u>New Section</u>. 422E.2 Imposition by County**. 1. A local sales and services tax shall be imposed by a county only after an election at which a majority of those voting on the question favors imposition. A local sales and services tax approved by a majority vote shall apply to all incorporated and unincorporated areas of that county. [On-Line]. (Available Netscape. Http: //www2.legis.state.ia.us/GA/77GA/Legislation/HF/02200/HF02282/ Current.html:)
Kansas	50 + 1	**§ 72.6761. General obligation bonds; purpose for issuance; when election required; contest of validity; limitations; temporary notes; tax levy.** When a board [of education] determines that it is necessary to purchase or improve a site or sites, or to acquire, construct, equip, furnish, repair, remodel or make additions to any building or buildings used for school district purposes, including housing and boarding pupils enrolled in an area vocational school operated under the board, or to purchase school buses, the board may submit to the electors of the unified district the question of issuing general obligation bonds for one or more of the above purposes, and upon the affirmative vote of the majority of those voting thereon, the board shall be authorized to issue the bonds. (Kansas Statutes Annotated. 1992. Volume No. 5A, p. 292.)

Kentucky	66⅔	**§ 162.080. Bond issues for school sites and buildings—Authorization—Election.** Whenever a board of education deems it necessary for the proper accommodation of the schools of its district to enlarge sites for school buildings, to purchase new sites, which in the case of independent districts may be not more than two (2) miles without the boundary lines of the district, to improve, remodel, or restore school buildings, to erect or equip new school buildings, or for any or all of these purposes, and the annual funds raised from other sources are not sufficient to accomplish the purpose, the board shall make a careful estimate of the amount of money required for the purpose shall determine the amount of money for which bonds shall be issued and the purpose to which the proceeds shall be applied. Upon request of the board of education of any district, the county clerk shall submit to the qualified voters of the district, the question as to whether bonds shall be issued for the purpose. The question shall be so framed that the voter may by his vote answer "for" or "against." **162.090. Issuance and sale of bonds—Proceeds—Tax to pay.** If two-thirds (⅔) of those voting on the question vote in favor of the proposition, the bonds shall be issued. The bonds shall be designated "school improvement bonds." (Kentucky Revised Statutes, Annotated, Official Edition. 1994. Volume 7A, pp. 341, 343.)
Louisiana	50 + 1	**§13. Funding; Apportionment.** (B) Minimum Foundation Program. The State Board of elementary and Secondary Education, or its successor, shall annually develop and adopt a formula which shall be used to determine the cost of a minimum foundation program of education in all public elementary and secondary schools as well as to equitably allocate the funds to parish and city school systems. Such formula shall proved for a contribution by every city and parish school system. . . . The legislature shall annually appropriate funds sufficient to fully fund the current cost to the state of such a program as determined by applying the approved formula in order to insure a minimum foundation of education in all public elementary and secondary schools. . . . The funds appropriated shall be equitably allocated to parish and city school systems. . . . Third: For giving additional support to public and secondary schools, any parish, school district, or subschool district, or any municipality or city school board which supports a separate city system of public schools may levy an ad valorem tax for a specific purpose, when authorized by a majority of the electors voting in the parish, municipality, district, or subdistrict in an election held for that purpose. The amount, duration, and purpose of the tax shall be in accord with any limitation imposed by the legislature. (Constitution of Louisiana. 1995. Revised December, 1995, pp. 76-77.)
Maine	50 + 1	**§ 1311. Power to borrow money. 2. Voter approval.** Bonds or notes for school construction purposes shall first be approved by a majority of voters of the district voting at an election called by the board of directors and held as provided in sections 1351 to 1354, except as is otherwise provided in this section. (Maine Revised Statutes Annotated. 1964. Volume 11, p. 277.)

Maryland	0 and 50 + 1 (50 + 1 only applies to school districts in the City of Baltimore and the County of Baltimore)	**[For school districts outside of the City of Baltimore and the County of Baltimore.] § 5-301. State payment of certain public school construction and capital improvements costs.** (a) Approved construction or capital improvement cost. – The Board of Public Works shall define by regulation what constitutes an approved public school construction or capital improvement cost. (b) State share. – The State shall pay the costs in excess of available federal funds of all public school construction projects and public school capital improvements in each county. The Board [of Public Works] is supreme in administration of school construction program. – In the administration of the public school construction program, the power of the Board of Public Works is plenary and supreme. **§ 5-302. Interagency Committee on School Construction.** The Interagency Committee on School Construction established by the Board of Public Works is a unit within the Department for administrative and budgetary purposes. (2) The Board of Public Works or the Interagency Committee on School Construction shall notify each county board and each local governing body of the annual allocation of school construction funds recommended to the Board of Public Works by the governor under the consolidated capital debt program of the State Finance and Procurement Article. The notification shall be made immediately after the Governor has recommended the allocations so that each county may structure its respective school construction and capital improvement priorities in accordance with the annual allocation and any amendments. (The Annotated code of the Public General Laws of Maryland. 1977 Replacement Volume. Education, pp. 161, 164-165.) **[For the City of Baltimore and the County of Baltimore.] Article XI. § 7. Debts and extension of credit.** From and after the adoption of this Constitution, no debt except as hereinafter provided in this section , shall be created . . . unless the debt or credit is authorized by an ordinance . . . and approved by a majority of the votes cast at that time and place. (The Annotated Code of the Public General Laws of Maryland. 1997. Constitutions, pp. 116-117.) **Article 25A, § 5. Enumeration. (P) (ii)** Any local law authorizing the borrowing of money or issuance of bonds or other evidences of indebtedness shall be submitted to the registered voters of the county for approval or rejection, if a petition for such submission is filed pursuant to the provisions of the charter and local laws of the county. If the charter contains no such provisions, any local law authorizing the borrowing of money or issuance of bonds or other evidences of indebtedness shall be submitted to the registered votes of the county for approval or rejection, if a petition for such submission, bearing the signatures of 10 percent centum or more of such voters, is filed with the board of supervisors of the county within 75 days after the enactment of such local law. (Annotated Code of the Public General Laws of Maryland. 1997. Volume 1, p. 531.)

Massachusetts	------	**§ 16. Regional School District; Body Politic and Corporate; Powers, etc.** A regional school district established under the provisions of the preceding section shall be a body politic and corporate with all the powers and duties conferred by law upon school committees, and with the following additional powers and duties: (d) To incur debt for the purpose of acquiring land and constructing, reconstructing, adding to, and equipping a school building or buildings for a term not exceeding twenty years . . . written notice of the amount of the debt and of the general purposes for which it was authorized shall be given to the board of selectmen in each of the towns comprising the district not later than seven days after the date on which said debt was authorized by the district committee; and no debt may be incurred until the expiration of sixty days from the date on which said debt was so authorized; and prior to the expiration of said period any member town of the regional school district may hold a town meeting for the purpose of expressing disapproval of the amount of debt authorized by the district committee, and if at such meeting a majority of the voters present and voting thereon express disapproval of the amount authorized by the district committee, the said debt shall not be incurred and the district school committee shall prepare another proposal which may be the same as any prior proposal and an authorization to incur debt therefor. (Annotated Laws of Massachusetts. 1997. Cumulative Supplement, pp. 13-14.)
Michigan	50 + 1	**§ 15.41351. Borrowing; bonds; issuance; purposes; amount limits; duration of indebtedness; exceptions; exclusion of refunding bonds, other bonds, from limitation; bonds subject to Municipal Finance Act; bonds or notes as full faith and credit tax limited obligations or voted and allocated tax levies; additional millage for payment without vote not permitted; ratification of bonds for textbooks purchase; confirmation, issuance. Sec. 1351. (1)** A school district may borrow money and issue bonds of the district to defray all or a part of the cost of purchasing, erecting, completing, remodeling, improving, furnishing, refurnishing, equipping, or reequipping school buildings, including library buildings, structures, athletic fields, playgrounds, or other facilities, or parts of or additions to those facilities; acquiring, preparing, developing, or improving sites, or parts of or additions to sites, for school buildings. **(2)** Except as otherwise provided . . . a school district shall not borrow money or issue bonds for a sum that, together with the total outstanding bonded indebtedness of the district, exceeds 5% of the state equalized valuation of the taxable property within the district, unless the proposition of borrowing the money or issuing the bonds is submitted to a vote of the school electors of the district at an annual or special election and approved by the **majority** of the school electors voting on the question. (Michigan Statutes Annotated. 1996. Volume 11A, pp. 717-718.)
Minnesota	50 + 1	**§ 124.2455. Bonds for certain capital facilities. (c)** A bond issue tentatively authorized by the board under this subdivision becomes finally authorized unless a petition signed by more than 15 percent of the registered voters of the school district is filed with the school board within 30 days of the board's adoption of a resolution stating the board's intention to issue bonds. The percentage is to be determined with reference to the number of registered voters in the school district on the last day before the petition is filed with the school board. The petition must call for a referendum on the question of whether to issue the bonds for the projects under this section. The approval of 50 percent plus one of those voting on the question is required to pass a referendum authorized by this section. (Minnesota Statutes Annotated. 1998. Volume 10A, p. 35.)

Mississippi	60	**§ 37-59-17. Determination of results of election; time period for issuance of bonds.** When the results of the election on the question of the issuance of such bonds shall have been canvassed by the election commissioners of such county or municipality, and certified by them to the school board of the school district, it shall by the duty of such school board to determine and adjudicate whether or not three-fifths (3/5) of the qualified electors who voted in such election in favor of the issuance of such bonds. Unless three-fifths (3/5) of the qualified electors who vote in such election vote in favor of the issuance of such bonds, then the school board of such school district shall issue such bonds, either in whole or in part, within two (2) years from the date of such election, or within two (2) years after the final favorable termination of any litigation affecting the issuance of such bonds, as such school board shall deem best. (Mississippi Code Annotated. 1972. Volume ten, pp. 697-698.)
Missouri	57 or 66⅔	**§ Section 26(b). Limitation on indebtedness of local government authorized by popular vote.** Any county, city, incorporated town or village or other political corporation or subdivision of the state, by vote of the qualified electors thereof voting thereon, may become indebted in an amount not to exceed five percent of the value of taxable tangible property therein as shown by the last completed assessment for state or county purposes, except that a school district by a vote of the qualified electors voting thereon may become indebted in an amount not to exceed ten percent of the value of such taxable tangible property. For elections referred to in this section the vote required shall be four-sevenths at the general municipal election day, primary or general elections and two-thirds at all other elections. (Missouri Constitution of 1875, Article X. Sec. 12; as adopted November 2, 1920; amended August 2, 1988, p. 101.)
Montana	50 + 1 or 60	**§ 20-9-428. Determination of approval or rejection of proposition at bond election.** (1) When the trustees canvass the vote of a school district bond election under the provisions of 20-20-415, they shall determine the approval or rejection of the school bond proposition in the following manner: (a) determine the total number of electors of the school district who are qualified to vote under the provisions of 20-20-301 from the list of electors supplied by the county registrar for such bond election; (b) determine the total number of qualified electors who voted at the last school bond election from the tally sheet or sheets for such election; (c) calculate the percentage of qualified electors voting at the school bond election by dividing the amount determined in subsection (1)(b) by the amount determined in subsection (1)(a); and (d) when the calculated percentage in subsection (1)(c) is 40% or more, the school bond proposition shall be deemed to have been approved and adopted if a majority of the votes have been cast in favor of such proposition, or otherwise it shall be deemed to have been rejected; or (e) when the calculated percentage in subsection (1)(c) is more than 30% but less than 40%, the school bond proposition shall be deemed to have been approved and adopted if 60% or more of the votes have been cast in favor of such proposition, otherwise it shall be deemed to have been rejected; or (f) when the calculated percentage in subsection (1)(c) is 30% or less, the school bond proposition shall be deemed to have been rejected. (Montana Code Annotated. 1997. Pp. 252-253.)

Nebraska	50 + 1	**§ 10-702. Issuance; election required; resubmission limited; submission at a statewide election; resolution; notice; counting boards.** The question of issuing school district bonds may be submitted at a special election or such question may be voted on at an election held in conjunction with the statewide primary or statewide general election. No bonds shall be issued until the question has been submitted to the qualified electors of the district and a majority of all the qualified electors voting on the question have voted in favor of issuing the same, at an election called for the purpose, upon notice given by the officers of the district at least twenty days prior to such election. (Revised Statutes of Nebraska Annotated. 1995. Volume 2, p. 598.)
Nevada	50 + 1	**§ 387.3287. Tax for account for replacement of capital assets or construction of new buildings for schools to accommodate community growth.** 1. Except as otherwise provided in subsections 4 and 5, upon the approval of a majority of the registered voters of a county voting upon the question, the board of county commissioners in each county may levy a separate tax pursuant to the provisions and subject to the limitations of subsections 1 and 2 of NRS 387.3285. **387.3285. Tax for fund for capital projects.** 1. Upon the approval of a majority of the registered voters of a county voting upon the question, the board of county commissioners in each county with a school district whose enrollment is fewer than 25,000 pupils may levy a tax which, when combined with any tax imposed pursuant to NRS 387.3287, is not more than 75 cents on each $100 of assessed valuation of taxable property within the county. The question submitted to the registered voters must include the period during which the tax will be levied. The period may not exceed 20 years. 2. Upon approval of a majority of the registered voters of a county voting upon the question, the board of county commissioners in each county with a school district whose enrollment is 25,000 pupils or more may levy a tax which, when combined with any tax imposed pursuant to NRS 387.3287, is not more than 50 cents on each $100 of assessed valuation of taxable property within the county. The question submitted to the registered voters must include the period during which the tax will be levied. The period may not exceed 20 years. (Nevada Revised Statutes. 1995. Volume 27, pp. 9756-9757.)
New Hampshire	66⅔	**§ 33:8. Town or District Bonds or Notes.** Except as otherwise specifically provided by law, the issue of bonds or notes by any municipal corporation, except a city, shall be authorized by a vote by ballot of ⅔, and the issue of tax anticipation notes, by a majority, of all the voters present and voting at an annual or special meeting of such corporation, called for the purpose; provided, however, that no such action taken at any special meeting shall be valid unless a majority of all the legal voters are present and vote thereat, unless the governing board of any municipality shall petition the superior court for permission to hold an emergency special meeting, which, if granted, shall give said special meeting the same authority as an annual meeting and provided further that the warrant for such special shall be published once in a newspaper having a general circulation. (New Hampshire Revised Statutes Annotated. 1955. 1988 Replacement Edition, p. 482.)

New Jersey	0 or 50 + 1	**Article 3. Appropriations. A. Type I Districts. § 18A:22-18. Capital projects; appropriations; estimation.** When a board of education of a type I district shall determine by resolution that it is necessary to sell bonds to raise money for any capital project authorized by law, it shall prepare and deliver to each member of the board of school estimate a statement of the ;amount of money estimated to be necessary for such purpose. **§18A:22-20. Capital projects; appropriations; how raised.** The governing body of the municipality shall, subject to the limitations hereinafter contained: borrow the sum or sums determined pursuant to N.J.S. 18A22-19, and secure repayment thereof, with interest thereon, at a rate not to exceed 6% per annum, by the authorization and issuance of bonds in the corporate name of the municipality, in accordance with the law. The governing body shall not be required to appropriate any amount which if added to the net school debt of the district at the date of the appropriation shall exceed 1½% of the average equalized valuations of taxable property as defined in section 18A:24-1, but may do so if it so determines by resolution. **B. Type II Districts. 18A:22-27. Type II districts with boards of school estimate; estimate by board of education; certification of estimate.** Whenever a board of education in a type II school district having a board of school estimate shall, by resolution adopted by recorded roll call affirmative vote of two thirds of its full membership, determine that it is necessary to sell bonds to raise money for any capital project or projects, itemizing such estimate so as to make it readily ;understandable, and the secretary of the board of education shall certify a copy of such resolution to each member of the board of school estimate of the district. **§ 18A:22-32. Type II districts without board of school estimate; determination of appropriation.** At or after the public hearing on the budget but not later than 18 days prior to the election, the board of education of each type II district having no board of school estimate shall fix and determine by a recorded roll call majority vote of its full membership the amount of money to be raised pursuant to section 5 of P.L. 1996, c. 138 (C. 18A:7F-5) and any additional amounts to be voted upon by the legal voters of the district at the annual election pursuant to section 5 of that act, which sum or sums shall be designated in the notice calling the election as required by law. **§ 18A:24-30. Submission and approval.** Whenever bonds are authorized to be issued by a type II school district under this chapter, the secretary of the board of education of the district shall transmit to the commissioner a certified copy of the bond proposal adopted by resolution of the board of education and approved by a majority of the legally qualified voters of the district voting on the proposal at an annual or special school election. (New Jersey Statutes Annotated. 1997. Title 18A, Education, 18A:20 to 18A:54D, pp. 16-30.)

New Mexico	50 + 1	**§ 22-18-8. Restriction on bond elections.** In the event a majority of those persons voting on a question submitted to the voters in a bond election vote against creating a debt by issuing general obligation bonds, no bond election shall be held on the same question for a period of two years from the date of the bond election, except upon the presentation of a petition pursuant to Section 22-18-2 NMSA 1978 and after expiration of at lest six months from the date of the previous bond election on the question. If a majority of those persons voting on a question submitted to the voters in a bond election for a second time within two years vote against creating a debt by issuing general obligation bonds, no bond election shall then be held on the same question for a period of two years from the date of first bond election on the question. (New Mexico Statutes Annotated. 1978. 1993 Replacement Pamphlet, p. 183.)
New York	50 + 1	**§ 416. School taxes and school bonds.** 1. A majority of the voters of any school district, present and voting at any annual or special district meeting, duly convened, may authorize such acts and vote such taxes as they shall deem expedient for making additions, alteration, repairs or improvements, to the sites or buildings belonging to the district, or for altering and equipping for library use any former schoolhouse belong to the district, or for purchase of other sites or buildings, or for a change of sites, or for the purchase of land and buildings for agricultural, athletic, playground or social center purposes, or for the erection of new buildings, or for building a bus garage, or for buying apparatus, implements, or fixtures, or for paying the wages of teachers, and the necessary expenses of the school, or for the purpose of paying any judgment, or for the payment or refunding of an outstanding bonded indebtedness, or for other purpose relating to the support and welfare of the school as they may, by resolution, approve. (Consolidated Laws of New York Annotated. 1988. Book 16, p. 496.)
North Carolina	50 + 1	**§ 159-61. Bond referenda; majority required; notice of referendum; form of ballot; canvass.** (a) If a bond order is to take effect upon approval of the voters, the affirmative vote or a majority of those who vote thereon shall be required. (b) The date of a bond referendum shall be fixed by the governing board, but shall not be more than one yea after adoption of the bond order. The governing board may call a special referendum for the purpose of voting on a bond issue on any day, including the day of any regular or special election held for another purpose (unless the law under which the bond referendum or other election is held specifically prohibits submission of other questions at the same time). A special bond referendum may not be held within 30 days before or days after a statewide primary, election, or referendum to be held in the same unit holding the bond referendum and already validly called or scheduled by law at the time at the bond referendum is called. In fixing the date of a bond referendum, the governing board shall consult the board of elections in order that the referendum shall not unduly interfere with other elections already scheduled or in process. Several bond orders or other matters may be voted upon at the same referendum. (Greater Statutes of North Carolina. 1994. Chapter 159, Local Government Finance, p. 63.)
North Dakota	60	**§ 21-03-07. Election required – Exceptions.** No municipality, and no governing board thereof, may issue bonds without being first authorized to do so by a vote equal to sixty percent of all the qualified voters of such municipality voting upon the question of such issue. (North Dakota Century Code. 1991. Volume 4A, p. 151.)

Ohio	50 + 1	**§ 3318.06. Resolution relative to tax levy in excess of ten-mill limitation; bond issue; submission to electors.** If a majority of those voting upon a proposition hereunder which includes the question of issuing bonds vote in favor thereof, and if the agreement provided for by section 3318.08 of the revised code has been entered into, the school district may proceed under Chapter 133. Of the Revised Code, with the issuance of bonds or bond anticipation notes in accordance with the terms of the agreement. (Ohio Revised Code Annotated. 1997. Title 33: Education—Libraries, p. 489.)
Oklahoma	60	**§ 15-103. Electors—Qualifications.** In case three-fifths (3/5) of the voters thereof voting at such election shall vote affirmatively for the issuance of said bonds, then the said board of education shall issue the same and not otherwise. The amount of bonds so voted upon and issued shall not cause the school district to become indebted in an amount, including existing indebtedness, in the aggregate exceeding five percent (5%) of the valuation of the taxable property therein, to be ascertained from the last assessment for state and county purposes previous to incurring of such indebtedness; but if the school district has an absolute need therefor, such district ;may, with the assent of three-fifths (3/5) of the voters thereof, voting at such election, incur indebtedness to an amount, including existing indebtedness, in the aggregate exceeding five percent (5%) but not exceeding ten percent (10%) of the valuation of the taxable property therein, to be ascertained from the last assessment for state and county purposes previous to the incurring of such indebtedness, for the purpose of acquiring or improving school sites, constructing, repairing, remodeling or equipping buildings or acquiring furniture, fixtures or equipment or more than one or all of the purposes; and such assent to such indebtedness shall be deemed to be sufficient showing of such absolute need. (Oklahoma Statutes Annotated. 1998. Title 70. Schools, pp. 556-557.)
Oregon	50 + 1	**§ 328.205 Power to contract bonded indebtedness; use of proceeds to pay expenses of issue.** (1) Common and union high school districts may contract a bonded indebtedness for any one or more of the following purposes in and for the district: (a) To acquire, construct, reconstruct, improve, repair, equip or furnish a school building or school buildings or additions thereto; (b) To fund or refund the removal or containment of asbestos substances in school buildings and for repairs made necessary by such removal or containment; (c) To acquire or to improve all property, real and personal, appurtenant thereto or connected therewith, including school buses; (d) To fund or refund outstanding indebtedness; and (e) To provide for the payment of debt. **§ 328.213 Issuance of negotiable interest-bearing warrants.** (1) When authorized by a majority of the electors of the [school] district, the board of a common or union high school district may contract a district debt for an amount which together with outstanding bonded indebtedness shall not exceed the bonding limit of the district as provided by ORS 328.245, for the purposes specified in ORS 328.205 and issue negotiable interest-bearing warrants of the district, evidencing such debt, and fix the time of payment of the warrants. Such warrants shall be considered a type of bond. (Oregon Revised Statutes. 1995. Volume 7, p. 33.)

Pennsylvania	50 + 1	**§ 7-701.1. Referendum or public hearing required prior to construction or lease.** Except where the approval of the electors is obtained to incur indebtedness to finance the construction of a school project, the board of school directors of any school district of the second, third or fourth classes, shall not construct, enter into a contract to construct or enter into a contract to lease a new school building or substantial addition to an existing school building without the consent of the electors obtained by referendum or without holding a public hearing as herein provided. In the event that a new school building or a substantial addition to an existing building is to be constructed or leased, the school board shall, by majority vote of all its members, authorize a maximum project cost and a maximum building construction cost to be financed by the district or authorized by lease rentals to be paid by the district. (Pennsylvania Statutes Annotated. 1992. Title 24 Education, p. 357.
Rhode Island	------	**§ 16-7-44. School housing, project costs.** School housing project costs, the date of completion of school housing projects, and the applicable amount of school housing project cost commitments shall be in accordance with the regulations of the commissioner of elementary and secondary education and the provisions of § § 16-7-35 to 16-7-47; provided, however, that school housing project costs shall include the purchase of sites, buildings, and equipment, the construction of buildings, and additions or renovations of existing buildings and/or facilities. Commencing with fiscal year 1990-1991, school housing project costs shall include the cost of interest payment on any bond issued after July 1, 1998. The board of regents for elementary and secondary education will promulgate rules and regulations for the administration of this section. These rules and regulations may provide for the use of lease revenue bonds, financial leases, capital reserve funding, and similar financial instruments to finance school housing provided that the net interest costs shall be less than what a general obligation bond would be, and further provided that the term of any bond, lease or similar instrument shall not be longer than the useful life of the project and such instruments are subject to the public review and voter approval otherwise required by law for the issuance of bonds, leases or similar instruments. Cities or towns issuing bonds, notes, leases or other evidences of indebtedness issued by municipal public buildings authority for the benefit of a local community pursuant to chapter 50 of title 45 shall not require voter approval. **§ 16-7-45. Annual appropriations.** The general assembly shall annually appropriate such sums as it may deem necessary to carry out the purposes of § § 16-7-35 to 16-7-47, and the state controller is hereby authorized and directed to draw his or her orders upon the general treasurer for the payment of the sum, or so much thereof as may be required from time to time, upon the receipt by the controller of properly authenticated vouchers. **§ 16-7-47. Addition to existing aid.** The provisions of § § 16-7-35 to 16-7-47 shall be in addition to any and all state aid for education as provided in any other general or special law. (General Laws of Rhode Island. 1956. Reenactment of 1996. Volume 3B, pp. 85-86.)

South Carolina	50 + 1	**§ 11-27-50. Effect of New Article X on bonds of school districts.** The Board of Trustees or other governing body (the governing body) of each of the school districts of the State shall be empowered to incur general obligation debt for their respective school districts as permitted by Section 15 of the New Article X and in accordance with its provisions and limitations. All laws relating to such matters shall continue in force and effect after the ratification date, but all such laws are amended as follows: 1. If no election be prescribed in such law and an election is required by New Article X, then in every such instance, a majority vote of the qualified electors of the school district voting in the referendum herein authorized is declared a condition precedent to the issuance of bonds pursuant to such law. The governing body of each of the school districts shall be empowered to order any such referendum as is required by New Article X or any other provisions of the Constitution, to prescribe the notice thereof and to conduct or cause to be conducted such referendum in the manner prescribed by Article 1, Chapter 71, title 59, Code of Laws of South Carolina, 1976. (Code of Laws of South Carolina Annotated. 1976. Volume 4A, Public Buildings and Property, p. 162.)
South Dakota	60 or 50 + 1	**§ 6-8B-2. Election required for issuance.** Unless otherwise provided, no bonds may be issued either for general or special purposes by any public body unless at an election sixty percent of voters of the public body voting upon the question vote in favor of issuing the bonds. The election shall be held in the manner described by law for other elections of the public body. **§ 6-3-1. Construction, furnishing, operation and maintenance of common building authorized—Use of existing building.** As used in §§ 6.3.1 to 6.3.8, inclusive, the term, political subdivision, means any county, any municipality, improvement district or any school district. Any two or more political subdivisions, each of which has a territory overlapping that of all of the others, may agree in the manner set forth in 6-3-2 to acquire a site for, purchase or construct, equip, furnish, operate and maintain a public building for their common use, for the purposes authorized by law in the case of each such subdivision, or to improve, extend, equip and furnish, for such purposes, any such building previously owned by the participating subdivisions or any of them. **§ 6.3.3. Appropriations and bonds authorized for common building—Bone issue prohibited until financing provisions complete.** The governing body of each participating political subdivision may appropriate money or may also issue the general obligation bonds of the subdivision, as provided in chapter 6-8B for the authorization, issuance and sale of bonds, for the payment of its share of the cost of the building or improvement; provided, that no bonds may be issued until provision has been made by each of the other participating subdivisions for the payment of their shares of the cost and a majority of all voters voting on the bond issue authorize it. (South Dakota Codified Laws. 1993 Revision. Volume 3A, pp. 9-10, 23.)

Tennessee	0 or 50 + 1	**§ 9-21-207. When an election is necessary.** – (a) No election upon a proposition for the issuance of general obligation bonds shall be necessary. (b) If a petition protesting the issuance of the general obligation bonds signed by at least ten percent (10%) of the registered voters of the local government, determined as of the date of publication of the notice required in § 9-21-206, or that portion, if applicable, which is liable to be taxed therefor, is filed with the official charged with maintaining the records of the local government within ten (10) days from the publication of the initial resolution, then no general obligation bonds shall be issued with the assent of the majority of the registered voters in the local government or a portion of the local government, if applicable, voting upon a proposition for the issuance of the general obligation bonds in the manner provided by §§ 9-21-209 and 9-21-210. **§ 9-21-209. Election resolution.** If it is necessary to hold an election on the proposition to issue general obligation bonds or if the governing body decides to hold an election to ascertain the will of the electorate even if no petition has been filed, then the election shall be called by the governing body of the local government. The governing body shall adopt a resolution (herein called the "election resolution) which shall supersede by its adoption, and immediately upon its adoption, the initial resolution, if any. (Tennessee Code Annotated. 1992. Volume 3A, pp. 156-157.)
Texas	50 + 1	**§ 45.003. Bond and Tax Elections.** (a) Bonds described by Section 45.001 may not be issued and taxes described by Section 45.001 or 45.002 may not be levied unless authorized by a majority of the qualified voters of the [school] district, voting at an election held for that purpose, at the expense of the district, in accordance with the Election Code, except as provided in this section. The resolution or order must state the date of the election, the proposition or propositions to be submitted and voted on, the polling place or places, and any other matters considered necessary or advisable by the governing board or commissioners court. (e) Before issuing bonds, a district must demonstrate to the attorney general with respect to the proposed issuance that the district has a projected ability to pay the principal of and interest on the proposed bonds and all previously issued bonds other than bonds authorized to be issued at an election held on or before April 1, 1991, and issued before September 1, 1992, from a tax at a rate not o exceed $0.05 per $100 of valuation. (Texas Codes Annotated. 1996. Education Code, p. 131.)
Utah	50 + 1	**§ 11-14-1. Municipality defined - Bond issues authorized - Purposes - Use of bond proceeds - Costs allowed.** (1) "Municipality," for the purpose of this chapter, includes cities, towns, counties, school districts, public transit districts, and improvement districts operating under the authority of Title 17A, Chapter 2, Part 3. **§ 11-14-2. Election on bond issues - Qualified electors - Resolution and notice.** The governing body of any municipality desiring to issue bonds under the authority granted in Section 11-14-1 shall by resolution provide for the holding of an election in the municipality on the question of the issuance of the bonds, and the bonds may be issued only if at the election the issuance of the shall have been approved by a majority of the qualified electors of the municipality who vote on the proposition. (Utah Code Annotated. 1997. Volume 5C, p.)

Vermont	50 + 1	**[School board goes to the voters during an annual or special meeting. Voting takes place at such meetings.]** § 1756. **Notice of meeting, authorization.** The clerk of the municipal corporation shall cause notice of such meeting to be published in a newspaper of know circulation in such municipality once a week for three consecutive weeks on the same day of the week, the last publication to be not less than five nor more than ten days before such meeting. Notice of such meeting shall also be posted in five public places within such municipality for two weeks immediately preceding such meeting. When a majority of all voters present and voting n the question at such meeting vote to authorize the issuance of bonds for said public improvements, the legislative branch shall be authorized to make such public improvements and issue bonds as hereinafter provided. [*However, if the school board opts to issue bonds without voter approval, a current statute could prohibit the sale of bonds*]. § 1755. **Submission to voters.** (a) On a petition signed by at least ten percent of the voters of a municipal corporation the proposition of incurring a bonded debt to pay for public improvements shall be submitted to the qualified voters thereof at any annul or special meeting to be held for that purpose, or, when the legislative branch of a municipal corporation at a regular or special meeting called for such purpose shall determine by resolution passed by a vote of a majority of those members present and voting, that the public interest or necessity demands improvements, and that the cost of the same will be too great to be paid out of the ordinary annual income and revenue, by vote of a majority of those members present and voting, it may order the submission of the proposition of incurring a bonded debt to pay for public improvements to the qualified voters of such municipal corporation at a meeting to be held for that purpose. The warning calling the meeting shall state the object and purpose for which the indebtedness is proposed to be incurred, the estimated cost of the improvements and the amount of bonds proposed to be issued, and shall fix the place where and the date on which the meeting shall be held and the hours opening and closing the polls. (Vermont Statues Annotated. 1992. Title 24, sections 1-3221, pp. 174-175.)
Virginia	50 + 1	§ **15.2.2611. Holding of election; order authorizing bonds; authority of governing body.** If a majority of the voters of the locality voting on the question approve the bond issue, the court shall enter an order to such effect, a copy of which shall be promptly certified by the clerk of the court to the governing body of the locality. The locality may then proceed to prepare, issue and sell its bonds up to the amount so authorized and in doing so shall have all the powers granted to the locality by this chapter with respect to incurring debt and issuing bonds. Bonds authorized by a referendum may not be used by a locality more than eight years after the date of the referendum; however, this eight-year period may, at the request of the governing body of the locality, be extended to up to ten years after the date of the referendum by order of the circuit court for the locality. (Code of Virginia Annotated. 1950. Volume 3A, 1997, Replacement Volume, p. 420.)

Washington	50 + 1 or 60	**§ 28A.530.020. Bond issuance—Election—Resolution to specify purposes.** (1) The question whether the bonds shall be issued, as provided in RCW 28A.530.010, shall be determined at an election to be held pursuant to RCW 39.36.050. If a majority of the votes cast at such election favor the issuance of such bonds, the board of directors must issue bonds: *Provided*, That if the amount of bonds to be issued, together with any outstanding indebtedness of the district that only needs a simple majority voter approval, exceeds three-eighths of one percent of the value of the taxable property in said district, as the term "value of the taxable property" is defined in RCW 39.36.015, then three-fifths of the votes cast at such election must be in favor of issuance of such bonds, before the board of directors is authorized to issue said bonds. (Revised Code of Washington Annotated. 1997. Common School Provisions, pp. 230-231.)
West Virginia	60	**§ 13-14. Bond issue proposal to be submitted to voters; election order.** No debt shall be contracted or bonds issued under this article until all questions connected with the same shall have been first submitted to a vote of the qualified electors of the political division for which the bonds are to be issued, and shall have received three fifths of all the votes cast for and against the same. The governing body of any political division referred to in this article may, and when requested so to do by a petition in writing, praying that bonds be issued and stating the purpose and amount thereof, signed by legal voters of the political division equal to twenty percent of the votes cast in a county or magisterial district for governor, or in a municipal district corporation or school district for mayor or members of the board of education, as the case may be, shall, by order entered of record, direct that an election be held for the purpose of submitting to the voters of the political division all questions connected with the contracting of debt and the issuing of bonds. Such order shall state: (m) In the case of school bonds, that such bonds, together with all existing bonded indebtedness, will not exceed in the aggregate five percent of the value of the taxable property in such school district ascertained in accordance with section 8, article X of the Constitution. (West Virginia Code Annotated. 1995. Volume 5, p. 533.)

Wisconsin	50 + 1	**§ 119.49. Bond issues.** (1)(a) If the [school] board deems it necessary to construct buildings or additions to buildings, to remodel buildings or to purchase school sites or to provide funds for any such purpose as a participant in a contract under s. 66.30(6), it may by a two-thirds vote of the members-elect send a communication to the common council of the city. (b) the communication shall state the amount needed under par[agraph] (a) and the purposes for which the funds will be used and shall request the common council to submit to the voters of the city at the next election held in the city the question of issuing school bonds in the amount and for the purposes stated in the communication. (2) Upon receipt of the communication, the common council shall cause the question of issuing such school bonds in the stated amount and for the stated school purposes to be submitted to the voters of the city at the next election held in the city. The question of issuing such school bonds shall be submitted upon a separate ballot or in some other manner so that the vote upon issuing such school bonds is taken separately from any other question submitted to the voters. If a majority of the electors voting on the school bond question favors issuing such school bonds, the common council shall cause the school bonds to be issued immediately or within the period permitted by law, in the amount requested by the board in the manner other bonds are issued. (3) The proper city officials shall sell or dispose of the bonds in the same manner as other bonds are disposed of. The entire proceeds of the sale of the bonds shall be placed in the city treasury, subject to the order of the board for the purposes named in the communication under sub. (1). Such school bonds shall be payable within 20 years from the date of their issue. (Wisconsin statutes Annotated. 1991. Sections 115.001 to 123.End, pp. 296-297.
Wyoming	50 + 1	**§ 22-21-110. Ballot canvass; results certified; effect of defeat.** Immediately after the closing of the polls, the election officials at each polling place shall proceed to canvass the ballots. The results disclosed by the canvass shall be certified by the election officials to the clerk of the political subdivision. If the majority of the ballots cast on a bond question is in favor of the issuance of the bonds, the proposal shall be approved, and the governing body of the political subdivision, in the manner provided by law, shall then proceed to declare the results of said election, and complete the printing, execution, advertising, and sale of bonds, but if the majority is opposed to such issuance, the proposal to issue bonds for the same general purpose shall not again be submitted to election within the same calendar year. (Wyoming Statutes Annotated. 1997. Volume 5, pp. 280-281.)

APPENDIX B

Checklist: Effective School Finance Campaigns

FOLLOWING is a checklist of preparations that should be completed by election day, whether the functions are undertaken by a school district or by independent citizen committees. Boards of education and school districts may be constrained by certain legal limitations. Independent citizen committees are permitted more leeway. School administrators should check with the board attorney to be certain that the board and school district do not violate state law.

- Develop a carefully planned strategy, based upon statistical analysis and critique of past campaigns.
- Canvass voters by independent citizen committee (house-to-house or by telephone).
- Citizens committee identify other "yes" voters from records and research (such as parents, staff members, recent grads, high school seniors, parents of incoming kindergartners, etc.).
- Conduct voter registration campaign.
- Set up volunteer committee to offer baby-sitting services and rides to the polls. Publicize these services appropriately.
- Complete arrangements for absentee ballots for the ill, elderly, out-of-town students and other members of the community who will be unable to make it to the polls.
- Citizens committee recruit a group of "poll watchers" recruited by the citizens committee for each polling place. (These are citizens committee volunteers—completely separate from the school district's election workers, who are paid and neutral.)

Source: Adapted from J. E. Swalm. 1989. "Turning No to Yes on Bond Issues." *School Leader*, 18(4): 23–26.

- Citizens committee arrange a telephone squad for election day.
- Set up and staff central election headquarters with knowledgeable resource persons, who are aware of the election laws and procedures.
- Arrange polling places (if possible, in school buildings).
- Arrange trained election workers.
- Check out equipment to make sure it is working properly.

ON ELECTION DAY

- Arrange for citizens committee campaign coverage at each poll.
- Assign a poll watcher at every polling place (who gets appropriate relief, lunch and breaks, and has access to the telephone, or is visited by a courier regularly during the day).
- Supply two sets of voter lists for every poll watcher (with the "yes" voters already carefully designated).
- Poll watchers check off the "yes" voters as they cast their ballots.
- About two hours before the polls close, pick up one set of lists and distribute to the telephone squad.
- The citizens committee's telephone squad then calls each "yes" voter who has not yet voted to urge participation in a positive way and to offer rides and baby-sitting.
- Due to the rush, telephone squad volunteers should be expected to contact no more than about 15 voters each.
- Prepare scripts or outlines in advance to make sure all messages are conveyed and calls move along quickly.
- The poll watchers complete the second list and return it later to the coordinator so that a finished record of voting is compiled for each voting district.
- Gather results at election headquarters, with all volunteers joining the get-together as they complete their assignments.
- Serve refreshments and encourage camaraderie.
- Have tally sheets ready for those who wish to "keep score" as the counts come in from the districts.
- Provide a chalkboard so that all can watch the results go up together.
- Plan for the news media.
- Invite reporters—or arrange to call them.

- Have a spokesperson available to answer questions or provide commentary to the news media.
- The school board and administration should provide staff with the results first thing the next day.

AFTER THE ELECTION

- Hold a debriefing. Allow a few days cooling off period first. Then analyze the campaign carefully and critically together, while it is still fresh in your minds.
- Identify and record your least useful approaches.
- Brainstorm new ideas.
- Conduct an informal random sampling of workers and voters. What did they think of your campaign?
- Compile all of your data on voter turnout, analyzing the results hour by hour and district by district. Draw conclusions.
- Compile a complete archival record of the election with all materials possible and appropriate commentary on each: brochures, news releases, bulletins, scripts, slide shows, flyers, clippings.
- Organize their materials and store them away carefully so that they can be used to plan next year's campaign.
- Express your thanks to all volunteers—preferably in a personal note.
- Begin to plan your basic schedule for the upcoming school finance campaign.

APPENDIX C

Planning in Debt Issuance

THE purpose of this exposition is to examine the nature of the planning process in debt issuance and to see what kinds of information school districts require when financing improvements.

School boards must plan and control their school district's operations. School boards exercise control over their operations by:

1. Selecting the course of action they wish to take (planning and decision making)
2. Issuing instructions and seeing that they are carried out (direction and supervision)
3. Adjusting their methods or their plans on the basis of analyses of the results achieved (responsive control)

Control always begins at the planning stage. No other technique controls the district's destiny as planning does. The difference between effective and ineffective planning can be so great as to overshadow the effect of all other control techniques combined.

Planning is the process of deciding on a course of action, of finding and choosing among the alternatives available. Planning takes three basic forms.

1. Policy formulation: Establishment of the major ground rules that determine the basic direction and shape of the district.
2. Decision making: The choice among alternative solutions to specific operating or financial problems within the prescribed policy limits.

Source: From *South Dakota School District Bonds: A Guide to Public Indebtedness* by Danforth, Meierhenry & Meierhenry, L.L.P. 1997. Sioux Falls, SD. Used by permission.

3. Periodic planning: The preparation of comprehensive operating and financial plans for specific intervals of time.

The relationships between the project and program planning and the periodic planning are complex. For example, the decision to improve and invest in the high school building and not improve and invest in the elementary buildings may be made in the process of periodic planning. However, both these kinds of decisions may also be made later, when changing conditions give evidence that this part of the periodic plan can no longer be carried out effectively (e.g., the school district has a great high school but the elementary schools fail to meet state standards).

Some school administrators and school board members do not consider the future in forming policy. Let us call these types of school districts the "short-term districts." The short-term district's policy formation often takes a simple "no tax, no fee" approach. The short-term district provides education on depreciating assets without providing for their replacement until the assets are unable to provide the educational services requested of them. This type of planning is referred to as "crisis planning," that is, no planning until there is a crisis.

The crisis often begins years earlier when predecessor boards do not plan for the future. School boards should ask themselves: What will the replacement or improvement cost? When determining the replacement or improvement of a school building, one must determine the useful life of existing assets of the school district and the estimated date of replacement or improvement. The following illustrates a simple future value of current replacement costs assuming a 4 percent inflation rate.

Asset	Remaining Life	Current Replacement Cost	Estimated Replacement Costs at End of Life
High School	25 Years	$4,500,000	$9,860,054
Bus	7 Years	40,000	52,637
East Grade School	15 Years	1,500,000	2,701,415

Most school boards do not think in terms of replacement until the replacement decision is upon them. Many times the school district has outstanding debt which limits the project. Therefore, in the analysis, a school board should also determine the school district's indebtedness and the limitations it will create on future boards. Debt payments use taxes and revenue, which may be needed for future improvements.

When a district plans to issue debt, it must analyze how the current debt issuance would affect all the district's assets. All assets' useful life, replacement cost, debt, and income should be studied to determine whether a crisis exists or is in the making. The school board should work within the political and economic realities to provide a policy and a plan that will see the district through to the next major improvement and beyond.

Glossary of Terms: Municipal Bonds

Accrued Interest: Interest earned on a bond since the last payment date.

Ad Valorem Tax: A tax based on the assessed value of property.

Amortization: Special periodic payments which pay off the debt.

Assessed Valuation: The valuation placed on property for purposes of taxation.

Basis Book: A book of mathematical tables used to convert yield percentages to equivalent dollar prices.

Basis Price: The price expressed in yield or net return on the investment.

Bond: An interest-bearing promise to pay with a specific maturity.

Callable Bond: A bond which is subject to redemption prior to maturity at the option of the issuer.

Closed Lien: A pledge made solely to one issue which prohibits further pledging of the resource.

Coupon: The part of a bond which evidences interest due. Coupons are detached from bonds by the holders usually semi-annually and presented for payment to the issuer's designated paying agent.

Coverage: This is a term usually connected with revenue bonds. It indicated the margin of safety for payment of debt service, reflecting the number of times or percentage by which earnings for a period of time exceed debt service payable in such period.

Source: From *South Dakota School District Bonds: A Guide to Public Indebtedness* by Danforth, Meierhenry & Meierhenry, L.L.P., 1997, Sioux Falls, SD. Used by permission.

Current Yield: A relation stated as a percentage of the annual interest to the actual market price of the land.

Debt Service: The statutory or constitutional maximum debt-incurring power of the school district.

Debt Ratio: The ratio of the issuer's debt to a measure of value, such as assessed valuation, real value, and so on.

Debt Service: Required payments for interest on and retirement of principal amount of a debt.

Default: Failure to pay principal or interest promptly when due.

Denomination: The face amount of par value of a bond which the issuer promises to pay on the bond's maturity date.

Discount: The amount, if any, by which the principal amount of the bonds exceeds the cost price.

Dollar Bond: A bond which is quoted and traded in dollars rather than in yield.

Double-Barreled Bonds: A bond secured by a pledge of two or more sources of payment (e.g., special assessments and unlimited taxing power of the issuer).

General Obligation: A bond secured by pledge of the issuer's full faith and credit and taxing power.

Gross Debt: The total of the debtor's obligation.

Interest: Compensation paid or to be paid for the use of money.

Interest Dates: The dates on which interest is payable to the holders of bonds, usually set at semi-annual intervals on the 1st or the 15th of the month.

Interest Rate: The interest payable each year, expressed as a percentage of the principal.

Issuer: A municipal unit which borrows money through sale of bonds.

Legal Opinion: An opinion concerning the legality of a bond issue by a recognized firm of municipal bond attorneys specializing in the approval of public borrowing.

Limited Tax Bond: A bond secured by the pledge of a tax which is limited as to rate or amount.

Marketability: The measure of ease with which a bond can be sold in the secondary market.

Maturity: The date upon which the principal of the bond becomes due and payable.

Net Debt: Gross debt less sinking fund accumulations and all self-supporting debt.

New Issue Market: Market for new issues of municipal bonds.

Official Statement or "O.S.": An official document prepared by the investment banker or the issuer which gives in detail the security and financial information relating to the issue.

Overlapping Debt: That portion of the debt of other governmental units for which residents of a particular school district are responsible.

Par value: The face amount of the bond usually $1,000 or $5,000.

Paying Agent: Place where the principal interest is payable, usually a designated bank or the treasurer's office of the issuer.

Premium: The amount, if any, by which the price exceeds the principal amount of the bond.

Principal: The face amount of a bond, exclusive of accrued interest.

Ratings: Designation used by investors' services to give relative indications of quality. Normally Moody or Standard & Poor are the rating agencies.

Refunding: A system by which a bond issue is redeemed from the proceeds of a new bond issue at conditions generally more favorable to the issuer.

Registered Bond: A bond whose ownership is registered with the issuer or its agents for principal and interest or for the principal only.

Revenue Bond: A bond payable from revenues secured from a project which pays its way by charging rentals to the users, such as toll bridges or highways, or from revenues from another source which are used for public purpose.

Secondary Market: Market for issues previously offered or sold.

Self-Supporting Debt: Debt incurred for a project or enterprise requiring no tax support other than the specific revenue earmarked for the purpose.

Serial Bond: A bond of an issue which has maturities scheduled annually or semi-annually over a period of years.

Sinking Fund: A reserve fund accumulated over a period of time for retirement of a debt.

Special Tax Bond: A bond secured by a special tax, such as gasoline or tax increment.

Subdivision: A unit of government, such as a county, city, school district, or town.

Tax Base: The total resources available for taxation.

Tax Exempt Bond: A bond, the interest on which is exempt from Federal income tax.

Tax Exempt Bond Fund: Registered unit of investment trusts, the assets of which are invested in diversified portfolio of interest-bearing municipal bonds issued by the states, cities, counties, and other political subdivisions.

Term Bond: A bond of an issue which has a single maturity.

Trading Market: The secondary market for issued bonds.

Trustee: A bank designated as the custodian of funds and official representative of bondholders.

Unlimited Tax Bond: A bond secured by pledge of taxes which may be levied in unlimited rate or amount.

Yield: The net annual percentage or income from an investment. The yield of a bond reflects interest rate, length of time to maturity, and write-off of premium or discounts.

AFTERWORD

AMERICAN schools face major challenges in their efforts to prepare students for the 21st century. Students still need to learn the basics of reading, writing, and arithmetic, but they also must be equipped to function in the Information Age and participate in the global economy. They deserve to be taught in safe, modern schools with the proper technology and resources, and research shows that they learn more when they are.

The American public knows that a strong public education system is the key to our future. We all understand that good schools are the cornerstones of healthy communities. While we have more students than ever, demands on local taxpayers are also growing. *School Bond Success: A Strategy for Building America's Schools* will be an invaluable reference for educators, administrators, parents, and local businesses as they work together with limited resources to rebuild and modernize their communities' schools.

Education is becoming more critical to the future of this nation, but in too many cases the national infrastructure for American schools is not up to the challenge. Too many of our school buildings are old and outdated. Many are downright unsafe. Paint is peeling. Roofs are leaking. Ceilings are falling in.

But that's not all. Record student enrollments are resulting in many overcrowded classrooms. Visit a school in a growing suburban area, and you're likely to see more than one trailer outside, a makeshift arrangement that, at best, is inadequate and can be intolerably hot in the warmer months and cold in the winter—especially in places like South Dakota.

Technology places even more demands on school facilities. Many older schools have only one or two outlets in a classroom—enough for an overhead projector, but not enough for a room full of computers. Some schools have only one phone line that comes into the building, and that goes into the principal's office. Upgrading electrical and phone lines can be a daunting proposition for many schools.

Members of Congress recognize that the states and local communities have the primary responsibility for educating children, but many of us believe this problem has grown so large, and the cost of failing so great, that the federal government must act. We have introduced legislation to provide federal assistance to communities that need to upgrade or expand their schools. Our objective is not to diminish local control in any way, but to enhance a community's ability to provide the best possible education for their children. I hope that our efforts, combined with the valuable contributions of Dr. Floyd Boschee, Dr. Carleton Holt, and others dedicated to American education, will help communities modernize our public school infrastructure to meet the challenges of the 21st century.

SENATOR TOM DASCHLE
United States Senate

INDEX

ABOUT THE AUTHORS

FLOYD BOSCHEE is a professor in the Division of Educational Administration, School of Education, University of South Dakota, where he teaches and conducts research in educational leadership, supervision, and curriculum. He also serves as a member of the board of education for the Vermillion School District in Vermillion, South Dakota. In eighteen years of public school service, he served as a teacher, coach, athletic director, and assistant superintendent for curriculum and instruction. He has also served as chairman of departments of education, published extensively in national journals, and authored or co-authored the books, *Grouping = Growth*, *Effective Reading Programs: The Administrator's Role*, *Outcome-Based Education: Developing Programs Through Strategic Planning*, *Authentic Assessment: The Key to Unlocking Student Success*, and *Special and Compensatory Programs: The Administrator's Role*.

CARLETON R. HOLT is a public school superintendent of the Dakota Valley School District in South Dakota. He has also served as superintendent of the Brandon Valley School District in South Dakota for thirteen years and as assistant to the superintendent at Webster City School District in Iowa for ten years. During his tenure as school superintendent, he guided three successful bond issue campaigns for construction of an elementary school, middle school, and high school. The victory margins ranged from 63 percent to 77 percent. He also served as band instructor and coach, published in national journals on bond elections, served as guest lecturer on successful marketing techniques for university classes and school board and administrative conferences, and consulted for local school districts in the United States and Canada.